It's another great book from CGP...

This book has everything you need to get your head around Science at KS3 (ages 11-14) — every topic is explained in a clear, straightforward style.

It's ideal if you're working at a higher level — it covers what would have been called Levels 5-7 in the old Curriculum.

CGP — still the best! ☺

Our sole aim here at CGP is to produce the highest quality books — carefully written, immaculately presented and dangerously close to being funny.

Then we work our socks off to get them out to you — at the cheapest possible prices.

Contents

Section 6 — Chemical Changes

Section 7 — The Earth and The Atmosphere

Section 8 — Energy and Matter

Section 9 — Forces and Motion

Section 10 — Waves

Section 11 — Electricity and Magnetism

Section 12 — The Earth and Beyond

Published by CGP

Editors:
David Maliphant, Matteo Orsini Jones, Rachael Rogers, Hayley Thompson, Sarah Williams

With thanks to Ian Francis, Jamie Sinclair and Karen Wells for the proofreading.

ISBN: 978 1 84146 230 1

Clipart from Corel®
Printed by Elanders Ltd, Newcastle upon Tyne.

The Microscope

A microscope is used for looking at objects that are <u>too small</u> to see with the <u>naked eye</u>. The <u>lenses</u> in the microscope <u>magnify</u> objects (make them <u>look bigger</u>) so that you can <u>see them</u>. It's magical stuff...

Learn the Different Parts of a Microscope

Here are some of the main parts of a <u>light microscope</u> — make sure you can <u>identify</u> them.

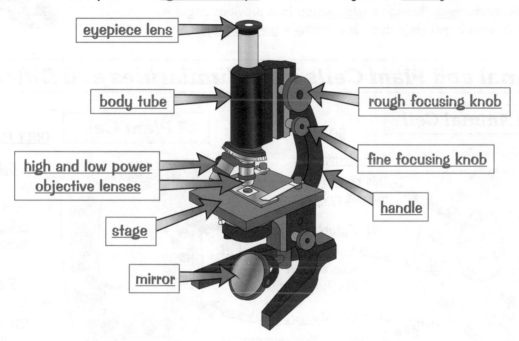

eyepiece lens

body tube

rough focusing knob

high and low power objective lenses

fine focusing knob

stage

handle

mirror

Follow These Easy Steps to Using a Light Microscope

1) Carry your microscope by the <u>handle</u>.

2) Place it near a <u>lamp</u> or a <u>window</u>, and angle the mirror so light shines up through the <u>hole</u> in the stage.

3) Clip a <u>slide</u> onto the <u>stage</u>. The <u>slide</u> should have the object(s) you want to look at <u>stuck to it</u>.

4) Select the <u>lowest</u> powered <u>objective lens</u>.

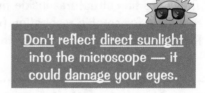

<u>Don't</u> reflect <u>direct sunlight</u> into the microscope — it could <u>damage</u> your eyes.

5) <u>Turn</u> the <u>rough focusing knob</u> to move the <u>objective lens</u> down to just above the slide.

6) <u>Look down</u> the <u>eyepiece lens</u> and <u>adjust the focus</u> using the <u>fine focusing knob</u>.

7) <u>Keep adjusting</u> until you get a <u>clear image</u> of whatever's on the slide.

DON'T BREAK THE SLIDE

<u>Always</u> turn the fine focusing knob so that the <u>objective lens</u> is moving <u>away</u> from the slide — so the lens and slide don't crash together.

8) If you need to see the slide with <u>greater magnification</u>, switch to a <u>higher powered objective lens</u> (a longer one).

9) Now refocus the microscope (repeat steps 5 to 7).

Microscopes — useful for looking at onions...

Teachers love getting you to look at <u>onions</u> under the microscope. Not a whole one mind, just the <u>slimy skin</u> between the onion layers (yuk). A microscope lets you see all the <u>tiny building blocks</u> (called <u>cells</u>) that the onion skin is made up of. It's, um, more interesting than it sounds. There's <u>more on cells</u> on the <u>next page</u>.

Cells

This page is about what living things are <u>made of</u>. Be prepared to find out that you're quite similar to a <u>plant</u>.

Living Things **are Made of** *Cells*

1) Another word for a <u>living thing</u> is an <u>organism</u>.
    <u>All organisms</u> are made up of <u>tiny building blocks</u> known as <u>cells</u>.

2) Cells can be <u>seen</u> through a <u>microscope</u> (see previous page) —
    but it helps if you <u>stain</u> them first (using a <u>coloured dye</u>).

Animal **and** *Plant Cells* **Have** *Similarities* **and** *Differences*

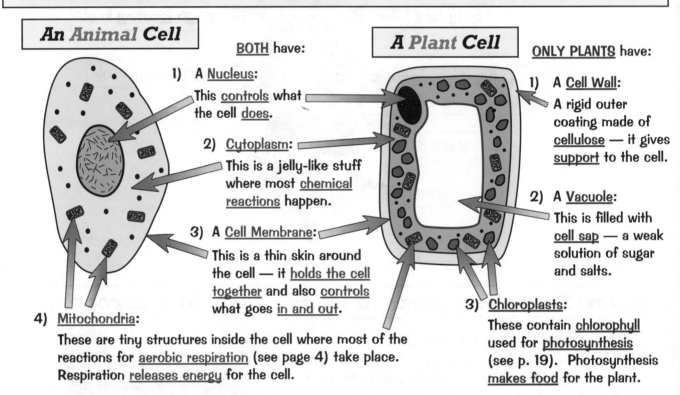

An Animal Cell

BOTH have:

1) A <u>Nucleus</u>:
    This <u>controls</u> what the cell <u>does</u>.

2) <u>Cytoplasm</u>:
    This is a jelly-like stuff where most <u>chemical reactions</u> happen.

3) A <u>Cell Membrane</u>:
    This is a thin skin around the cell — it <u>holds the cell together</u> and also <u>controls</u> what goes <u>in and out</u>.

4) <u>Mitochondria</u>:
    These are tiny structures inside the cell where most of the reactions for <u>aerobic respiration</u> (see page 4) take place. Respiration <u>releases energy</u> for the cell.

A Plant Cell

ONLY PLANTS have:

1) A <u>Cell Wall</u>:
    A rigid outer coating made of <u>cellulose</u> — it gives <u>support</u> to the cell.

2) A <u>Vacuole</u>:
    This is filled with <u>cell sap</u> — a weak solution of sugar and salts.

3) <u>Chloroplasts</u>:
    These contain <u>chlorophyll</u> used for <u>photosynthesis</u> (see p. 19). Photosynthesis <u>makes food</u> for the plant.

Some Living Things **are** *Unicellular*

1) <u>Animals</u> and <u>plants</u> are made up of <u>lots of cells</u>. They're <u>multicellular</u> organisms.

2) But many living things are made up of <u>only one cell</u> — these are called <u>unicellular</u> organisms.
    Unicellular organisms have <u>adaptations</u> to help them <u>survive</u> in the environment they live in, e.g.

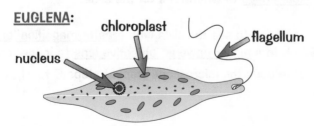

EUGLENA:

chloroplast

flagellum

nucleus

Euglena live in <u>water</u>. They have a <u>tail-like structure</u> called a <u>flagellum</u> to help them <u>swim</u>.

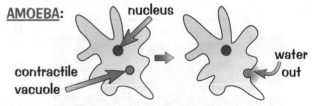

AMOEBA:

nucleus

contractile vacuole

water out

Some amoeba also live in <u>water</u>. They use a <u>contractile vacuole</u> to collect any <u>excess water</u> inside them and <u>squeeze it out</u> at the cell membrane.

Cells — they're great for locking things up...

You need to learn what all the bits and pieces in animal and plant cells <u>look like</u>, so you can <u>identify</u> them under a <u>microscope</u>. Have a go at <u>drawing</u> the top two cells above for yourself and then <u>labelling</u> them.

Cell Organisation

Cells are <u>organised</u>. They write all their appointments down in a little diary and never forget your birthday.

Learn *How Cells* are *Organised*

In organisms with <u>lots of cells</u> (like <u>animals</u> and <u>plants</u>), the cells are <u>organised</u> into <u>groups</u>. Here's how:

A group of <u>similar cells</u> come together to make a <u>tissue</u>.
A group of <u>different tissues</u> work together to make an <u>organ</u>.
A <u>group of organs</u> work together to make an <u>organ system</u>.
A multicellular <u>organism</u> is usually made up of <u>several organ systems</u>.

Here's a rather jolly example from a <u>plant</u>.
Don't forget that the sequence applies just as well to <u>animals</u>.

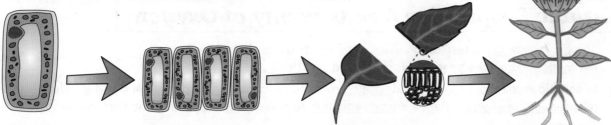

palisade <u>CELLS</u>...   ...make up palisade <u>TISSUE</u>...   ...which, with other tissues, makes up a leaf (an <u>ORGAN</u>)...   ...which, with more leaves and other organs, makes up the shoot system (an <u>ORGAN SYSTEM</u>). Various organ systems make up a full plant (an <u>ORGANISM</u>).

A palisade cell is just the name for a particular type of plant cell.

Stuff *Moves Into* and *Out of* *Cells by* Diffusion

1) Cells need things like <u>glucose</u> (a sugar) and <u>oxygen</u> to <u>survive</u>. They also need to <u>get rid</u> of <u>waste products</u>, like <u>carbon dioxide</u>.

2) These materials all <u>move into</u> or <u>out of cells</u> by a process called <u>diffusion</u>.

3) Diffusion is where a substance <u>moves</u> from an area of <u>high concentration</u> (where there's <u>lots of it</u>) to an area of <u>low concentration</u> (where there's <u>less of it</u>) — just like glucose in this diagram...

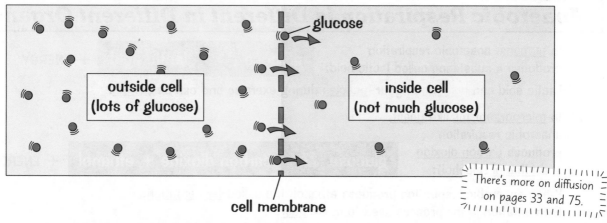

outside cell (lots of glucose)   glucose   inside cell (not much glucose)

cell membrane

There's more on diffusion on pages 33 and 75.

Get organised — learn this page...

Remember: cells ⟹ tissues ⟹ organs ⟹ organ systems ⟹ organisms. You need to get your head around <u>diffusion</u> too — it comes up <u>all the time</u> in KS3 science, so it's worth getting to grips with now.

Respiration

As <u>life processes</u> go, this is the big one — <u>respiration</u>. Without it, you'd be well and truly <u>stuck</u>.

Respiration is a Chemical Reaction

1) Respiration happens in <u>every cell</u> of <u>every living organism</u>.

2) Respiration is the process of <u>releasing energy</u> from <u>glucose</u> (a sugar).

3) The energy released by respiration is used for <u>all the other chemical reactions</u> that keep you <u>alive</u>. For example, the reactions involved in:

<u>building proteins</u>

<u>muscle contraction</u>

<u>keeping warm</u>

Aerobic Respiration Needs Plenty of Oxygen

1) <u>Aerobic respiration</u> is respiration using <u>oxygen</u>. It takes place in the <u>mitochondria</u> (see page 2) of <u>animal</u> and <u>plant cells</u>.

2) In aerobic respiration, <u>glucose</u> and <u>oxygen</u> react to produce <u>carbon dioxide</u> and <u>water</u>. This reaction releases <u>lots of energy</u>. Here's a <u>word equation</u> to show what happens in the reaction — <u>learn it</u>:

glucose + oxygen $\Longrightarrow$ carbon dioxide + water **+ ENERGY**

These are the <u>reactants</u>.   These are the <u>products</u>.

There's more on chemical reactions and word equations on pages 48 and 51.

Anaerobic Respiration Takes Place Without Oxygen

1) <u>Anaerobic respiration</u> is respiration <u>without oxygen</u>.

2) Anaerobic respiration is <u>less efficient</u> than aerobic respiration, so it releases <u>less energy</u>.

3) Because of this, anaerobic respiration usually only happens when cells <u>can't get enough oxygen</u>.

E.g. when you <u>exercise</u>, your body can't always get enough oxygen to your <u>muscle cells</u>. So they start to respire <u>anaerobically</u>.

Anaerobic Respiration is Different in Different Organisms

1) In <u>humans</u>, anaerobic respiration produces a substance called lactic acid:

glucose $\Longrightarrow$ lactic acid **+ ENERGY**

Lactic acid can <u>build up</u> in your <u>muscles</u> during exercise and can be <u>painful</u>.

2) In <u>microorganisms</u> like <u>yeast</u>, anaerobic respiration produces <u>carbon dioxide</u> and <u>ethanol</u> (alcohol):

glucose $\Longrightarrow$ carbon dioxide + ethanol **+ ENERGY**

When anaerobic respiration produces ethanol, it's called <u>fermentation</u>. Fermentation is the process used to make <u>beer</u>.

Respiration — it's energetic stuff, innit...

It can be quite tricky to get your head around the idea of respiration, but it just means releasing <u>energy</u> from <u>glucose</u>. <u>Aerobic</u> respiration does this with <u>oxygen</u>, while <u>anaerobic</u> respiration does it <u>without</u>. Sorted.

Section Summary

Welcome to your very first Section Summary. It's full of questions written especially for finding out what you actually know — and, more importantly, what you don't. Here's what you have to do... 1) Go through the whole lot of these Section Summary questions and try to answer them. 2) Look up the answers to any you can't do and try to really learn them (hint: the answers are all somewhere in Section 1). 3) Try all the questions again to see if you can answer more than you could before. 4) Keep going till you get them all right.

1) What part of a microscope do you clip your slide onto?

2) What do the focusing knobs on a microscope do?

3) Why should you always move the objective lens away from the slide when you're focusing a microscope?

4) What is an organism?

5) What instrument would you use to look at a cell? What can you do to help see cells?

6) Name four parts that both plant cells and animal cells have. Say what they all do.

7) Name three parts of a cell that only plant cells have.

8) Give an example of a unicellular organism and say how it's adapted to its environment.

9) Explain the meaning of: a) tissue   b) organ c) organ system. Give an example of each.

10) What is diffusion?

11) Give two examples of substances that move into or out of cells by diffusion.

12) What's the name of the process that goes on in every cell, releasing energy?

13) What is the energy released by this process used for? Give three examples.

14) What is aerobic respiration? Where does it take place in plant and animal cells?

15) Write down the word equation for aerobic respiration.

16) Give two differences between aerobic respiration and anaerobic respiration in humans. When might humans respire anaerobically?

17) What are the products of anaerobic respiration in yeast?

18) What is fermentation? What can fermentation be used to make?

Nutrition

Nutrition is <u>what you eat</u> — and what you eat is really <u>important</u> for your <u>health</u>.
A <u>balanced diet</u> will have the right amount of the <u>five nutrients</u> listed below, as well as <u>fibre</u> and <u>water</u>.

1) Carbohydrates

Starch   Contained in   Bread / potatoes / cereals   Used for   Energy

These are like <u>fuel</u> for your body. <u>Active</u> or <u>growing</u> folk need <u>lots</u> of <u>carbohydrate</u>.

2) Proteins

Proteins   Contained in   Meat / eggs / fish   Used for   Building Cells

<u>Proteins</u> are vital for <u>growth</u> and to <u>repair</u> damaged areas.

3) Lipids (Fats and Oils)

Fat   Contained in   Butter / cooking oil / cream   Used for   Energy

<u>Lipids</u> act as a <u>store of energy</u> — which you use if your body <u>runs out</u> of <u>carbohydrates</u>.

4) Vitamins

E.g. A B$_1$ B$_2$
C D E   Contained in   Vegetables / fruit / cereals   Used for

<u>Vitamins</u> are only needed in <u>very small amounts</u> — they keep many <u>vital processes</u> happening.

5) Minerals

	Iron	Calcium	Sodium
E.g. —			
Found in —			
Needed for –	Blood	Teeth/Bones	Nerves

Fibre

Contained in   Vegetables / fruit / cereals   Used for

<u>Fibre</u> helps <u>food move</u> through your <u>digestive system</u>.

Water

Contained in   Used for

About <u>75%</u> of your body is <u>water</u> and all <u>chemical reactions</u> (e.g. digestion), take place in water – it's <u>well</u> important!

Obviously we only had gruel when I was a lad... sob ...sob...

Five types of <u>nutrient</u>, what you <u>find them in</u> and what they're <u>for</u>. And then there's <u>fibre</u> and <u>water</u> too.
<u>Learn</u> each section and test yourself with the <u>only method</u> that works — <u>covering the page</u> and <u>jotting it down</u>.

More on Nutrition

Your body needs energy <u>all the time</u>. Even when you're asleep your body is using energy just to <u>keep you alive</u>. You get the energy your body needs from <u>carbohydrates</u> and <u>fats</u> in your <u>diet</u>.

Different People Have Different Energy Requirements

1) The <u>amount of energy</u> you need each day depends on your <u>body mass</u> ("weight") and level of <u>activity</u>.

2) The <u>heavier</u> and the <u>more active</u> you are, the <u>more energy</u> you will need.

You Can Work Out Your Daily Basic Energy Requirement...

1) Every <u>cell</u> (see page 2) in the body needs <u>energy</u>. So the <u>bigger</u> you are, the <u>more cells</u> you have, and the more energy you'll need.

2) You also need <u>energy</u> to <u>move</u>, and it takes <u>more</u> energy to move a <u>bigger mass</u>.

3) For every <u>kg</u> of <u>body mass</u>, you need <u>5.4 kJ</u> of energy every <u>hour</u>. This is the <u>basic energy requirement (BER)</u> needed to maintain <u>essential</u> bodily functions.

A kJ is a unit of energy.

You calculate it like this: → **Daily BER (kJ/day) = 5.4 × 24 hours × body mass (kg)**

E.g. a 60 kg person requires 5.4 × 24 × 60 = 7776 kJ/day

...and How Much Extra Energy You Need for Your Activities

1) The <u>more active</u> you are the <u>more energy</u> you will need.

2) For example, a 60 kg person will use about <u>400 kJ walking</u> for half an hour but <u>1500 kJ running</u> for half an hour.

3) To find out how much <u>energy you need in a day</u> you have to <u>add together</u> your <u>daily BER</u> and the <u>extra energy</u> you use in your <u>activities</u>.

400 kJ   1500 kJ

An Unbalanced Diet Can Cause Health Problems

Obesity

1) If you <u>take in more energy</u> from your diet <u>than you use</u>, your body will store the <u>extra energy</u> as <u>fat</u> — so you will <u>put on weight</u>.

2) If you weigh <u>over 20% more</u> than the recommended weight for your height, then you are classed as <u>obese</u>.

3) Obesity can lead to <u>health problems</u> such as <u>high blood pressure</u> and <u>heart disease</u>.

Starvation and Deficiency Diseases

1) Some people don't get <u>enough food to eat</u> — this is <u>starvation</u>.

2) The effects of starvation include <u>slow growth</u> (in children), being <u>more likely</u> to get <u>infections</u>, and <u>irregular periods</u> in women.

3) Some people don't get enough <u>vitamins or minerals</u> — this can cause <u>deficiency diseases</u>. For example, a lack of <u>vitamin C</u> can cause <u>scurvy</u>, a deficiency disease that causes problems with the skin, joints and gums.

Summon up the energy to learn this page...

You should be able to <u>work out</u> how much <u>energy</u> a person needs in a day. Also, make sure you understand about the <u>health problems</u> that can be caused from eating <u>too much</u> or <u>too little</u> food.

8

Digestion

Digestion's fab. The body breaks down almost all the food we shove into our mouths, so we can use the nutrients it contains. But it's not easy — lots of different organs have to work together to get the job done.

Digestion *is All About* Breaking Down Food

There are two steps to this. The first is quick, the second isn't:

1) Breaking down the food MECHANICALLY, e.g. chewing with teeth:
2) Breaking down the food CHEMICALLY — with the help of proteins called enzymes. Enzymes are biological catalysts — this means they speed up the rate of chemical reactions in the body.

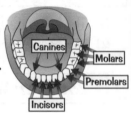

Eight Bits *of* The Alimentary Canal

1) Mouth

Digestion starts here where the teeth have a good old chew and mix the food with saliva. Saliva contains an enzyme (called amylase) that breaks down carbohydrates.

2) Oesophagus

(Food pipe) Links the mouth to the stomach.

3) Stomach

1) Here the food mixes with protease enzymes which digest proteins. The stomach contains muscular tissue to move the stomach wall and churn up food.

2) Hydrochloric acid is present to kill harmful bacteria and give a low pH for the enzymes to work.

4) Liver

The liver makes bile, which breaks fats into tiny droplets (emulsification). It's also alkaline to give the right pH for the enzymes in the small intestine.

5) Pancreas

The pancreas contains glandular tissue, which makes three enzymes:

1) PROTease digests PROTein.
2) CARBOHYDRAse digests CARBOHYDRAtes.
3) LIPase digests LIPids — i.e. fats.

7) Large intestine

Here water is absorbed — so we don't all shrivel up.

8) Rectum

Food usually contains some materials that we can't digest. This undigested food is stored as faeces. Here the digestion story ends when it plops out of the anus — egestion.

6) Small intestine

1) This produces more enzymes to further digest proteins, carbohydrates and fats.
2) Food is also absorbed through the gut wall into the blood, which then takes it around the body to wherever it's needed.

The alimentary canal — don't try to float a boat down it...

As a first step you need to memorise the headings — including each of the eight bits of the alimentary canal. Then there's just the small matter of learning all the details that go with them. Get an image in your head of the whole page and gradually learn the details as you scribble on. It'll get easier the more you practise.

More on Digestion

Well <u>would you believe it</u>? There's more to learn about digestion.

Absorption of Food Molecules

1) <u>Big</u>, <u>insoluble</u> food molecules <u>can't</u> pass through the <u>gut wall</u>.

2) So enzymes are used to <u>break up</u> the big molecules into <u>smaller</u>, <u>soluble ones</u>.

3) These small molecules <u>can</u> pass through the <u>gut wall</u> into the <u>blood</u>.
   They are then carried round the body, before
   passing into <u>cells</u> where they are used.

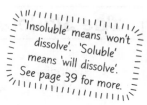

'Insoluble' means 'won't dissolve'. 'Soluble' means 'will dissolve'. See page 39 for more.

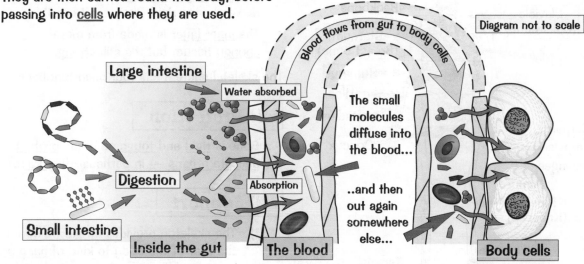

The Small Intestine is Covered with Millions of Villi

1) Food molecules are <u>absorbed into the blood</u> in the <u>small intestine</u>.

2) The small intestine is lined with tiny <u>finger-like projections</u> called <u>VILLI</u>.

3) Villi are <u>perfect</u> for <u>absorbing food</u> because:

   • They have a <u>thin outer layer of cells</u>.

   • They have a <u>good blood supply</u>.

   • They provide a <u>large surface area</u> for absorption.

Villi is the plural of villus — i.e. it's one villus but two (or more) villi.

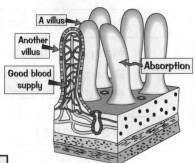

Bacteria are Really Important in the Gut

1) Bacteria are <u>unicellular organisms</u> (see page 2).

2) There are about <u>100 trillion bacterial cells</u> in the <u>alimentary canal</u>. That's <u>loads</u>.

3) Most of these are in the <u>end part</u> of the <u>small intestine</u> and in the <u>large intestine</u>.

4) Some types of bacteria can make you really <u>ill</u> if they get into your body,
   but the <u>bacteria</u> found <u>naturally in your gut</u> actually do a lot of <u>good</u>:

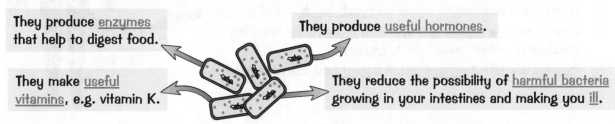

They produce <u>enzymes</u> that help to digest food.

They produce <u>useful hormones</u>.

They make <u>useful vitamins</u>, e.g. vitamin K.

They reduce the possibility of <u>harmful bacteria</u> growing in your intestines and making you <u>ill</u>.

Just like convicts, food can't pass through walls...

As well as just looking pretty, the <u>diagrams</u> on this page are really important for helping you understand how food is <u>broken down</u> and <u>absorbed</u> — so make sure you look at them properly and absorb the information. :)

The Skeleton and Muscles

The human skeletal system is made up of <u>206 bones</u> and the muscular system has around <u>640 muscles</u>. Together these systems are really important for allowing you to <u>move around</u>.

The Skeletal System

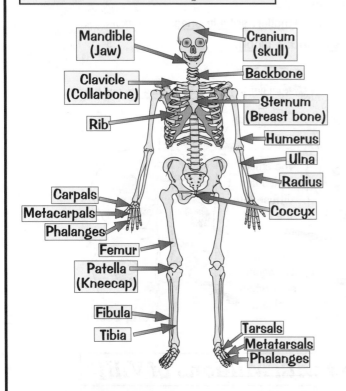

Mandible (Jaw)

Cranium (skull)

Clavicle (Collarbone)

Backbone

Rib

Sternum (Breast bone)

Humerus

Ulna

Radius

Carpals

Metacarpals

Phalanges

Coccyx

Femur

Patella (Kneecap)

Fibula

Tibia

Tarsals

Metatarsals

Phalanges

Bones are made from <u>different types of tissue</u>:

* The <u>outer layer</u> of bone is made from <u>really strong and hard</u> tissue — this makes bones <u>rigid</u> (they can't bend).
* The <u>inner layer</u> is made from more <u>spongy</u> tissue, but it's still <u>strong</u>.

The skeletal system has <u>four</u> main functions:

1) Protection

Bone is <u>rigid</u> and <u>tough</u> so it can <u>protect delicate organs</u> — in particular the <u>brain</u>.

2) Support

1) The skeleton provides a <u>rigid frame</u> for the rest of the body to kind of <u>hang off</u> — kind of like a custom made coat-hanger.
2) All the <u>soft tissues</u> are <u>supported</u> by the skeleton — this allows us to <u>stand up</u>.

3) Production of Blood Cells

1) Many bones have a soft tissue called <u>bone marrow</u> in the <u>middle</u> of them.
2) Bone marrow produces <u>red blood cells</u> (which carry <u>oxygen</u> around the body) and <u>white blood cells</u> (which help to <u>protect</u> the body from <u>infection</u>).

4) Movement

1) <u>Muscles</u> are <u>attached</u> to bones (see below).
2) The action of muscles allows the skeleton to <u>move</u>.
3) <u>Joints</u> (e.g. the knees and elbows) also allow the skeleton to move.

The Muscular System

1) <u>Muscles</u> are attached to bones via <u>tough bands</u> called <u>tendons</u>.
2) When a <u>muscle contracts</u> it applies a <u>force</u> to the bone it's attached to, which makes the <u>bone move</u>.
3) <u>Muscles</u> are found <u>in pairs</u> round a <u>joint</u> (see the next page).

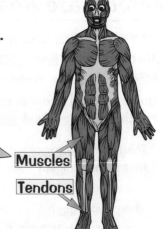

Muscles

Tendons

A body without bones? Ever seen a tent without poles...

Lots of bone-tinglingly exciting facts to <u>learn</u> here. You don't need to learn the names of all those bones — <u>much more important</u> than the names are what the bones and muscles <u>actually do</u>. The skeletal system has <u>four main functions</u> — you know what to do, cover up the page and see if you can scribble them all down.

How Muscles Work

Antagonistic Muscles *Work in* Pairs

1) Antagonistic muscles are pairs of muscles that work against each other.

2) One muscle contracts (shortens) while the other one relaxes (lengthens) and vice versa.

3) They are attached to bones with tendons. This allows them to pull on the bone, which then acts like a lever (see below).

4) One muscle pulls the bone in one direction and the other pulls it in the opposite direction — causing movement at the joint.

5) The biceps and triceps muscles in the arm are examples of antagonistic muscles. So are the hamstrings and quadriceps in the legs.

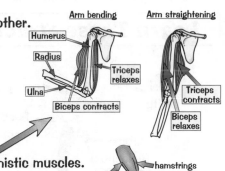

You Can *Measure the* Force *Applied by a* Muscle

Let's look at a muscle in the arm as an example:

> The study of forces acting on the body is called biomechanics.

Start By *Calculating the* Moment

1) A pivot is the point around which a rotation happens. A lever is a bar attached to a pivot.

2) When a force acts on something that has a pivot, it creates a "turning effect" known as a moment (see page 82).

3) The arm works as a lever with the elbow as a pivot. This means when a force acts on the arm there's a moment.

> 'Perpendicular distance' is the distance at a right angle from the pivot to the line of force.

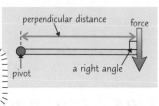

4) To calculate the size of a moment, you can use this equation:

$$\text{Moment} = \text{force} \times \text{perpendicular distance}$$

In newton metres (Nm)　In newtons (N)　In metres (m)

5) In the diagram here, the weight (a force) in the hand is creating a moment.

6) The weight has a force of 12 N. It is 0.3 m away from the pivot (the elbow). So using the equation above, the moment of the weight is 12 × 0.3 = 3.6 Nm.

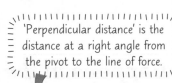

7) But the weight is not the only thing applying a force to the arm — the muscle is applying a force to counteract the moment of the weight and keep the arm still. For the arm to stay still, the moment of the muscle has to be the same as the moment of the weight (but acting in the opposite direction).

Now *Work Out The* Force *Applied By the* Muscle

You can rearrange the equation above to calculate the force applied by the muscle:

In the example above, the weight has a moment of 3.6 Nm, so the muscle must also have a moment of 3.6 Nm.

$$\text{Force} = \text{moment} \div \text{perpendicular distance}$$

In newtons (N)　In newton metres (Nm)　In metres (m)

The distance between the muscle and the pivot (elbow) is 0.05 m. So the force applied by the muscle is 3.6 ÷ 0.05 = 72 N.

Hang on a moment... what?

All this talk of forces and levers and moments can be pretty tricky to get your head around. But stick with it — you'll really impress your teacher with your super science knowledge if you can explain how muscles work.

Gas Exchange

You need to get oxygen from the air into your bloodstream. You also need to get rid of the carbon dioxide that's in your bloodstream. Funnily enough, this exchange of gases happens in your gas exchange system.

Learn These Structures in the Gas Exchange System

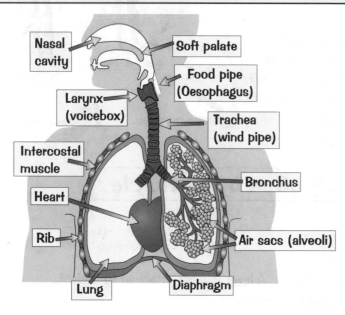

1) The lungs are like big pink sponges. They're protected by the ribcage.

2) The diaphragm is a muscle that sits underneath the ribcage. It moves up when it relaxes and down when it contracts. This movement helps to get air in and out of your lungs (see next page).

3) The air that you breathe in goes through the trachea. This splits into two tubes called 'bronchi' (each one is 'a bronchus'), one going to each lung.

4) The bronchi split into smaller tubes called bronchioles.

5) The bronchioles end at small air sacs in the lungs called alveoli. These are where gas exchange takes place.

Gas Exchange Happens in the Lungs

1) Air is inhaled into the lungs.

2) Some of the oxygen in the inhaled air passes into the bloodstream to be used in respiration (see page 4).

3) Carbon dioxide is a waste product of respiration. In the lungs it passes out of the blood and is then breathed out.

4) The gases pass into or out of the bloodstream by diffusion — where a substance moves from where there's lots of it to where there's less of it — see page 3.

5) The lungs are well adapted for gas exchange:

    1) They're moist.
    2) They have a good blood supply.
    3) The alveoli (air sacs) give the lungs a big inside surface area.

I love ribs — spare ones are my favourite though...

There are a couple of quite detailed diagrams here which need learning. Sooner or later you're expected to learn all the structures in the gas exchange system and what they do, so you may as well start now.

Breathing

Breathing is how the air gets <u>in and out</u> of your <u>lungs</u>. It's definitely a useful skill.

The Mechanism of Breathing

The <u>Bell Jar</u> demonstration shows us what's <u>going on</u> when you <u>breathe</u>:

1) First you pull the rubber sheet <u>down</u> — like it's your <u>diaphragm</u>.

2) This <u>increases</u> the <u>volume</u> inside the bell jar, which <u>decreases</u> the <u>pressure</u>.

3) The drop in pressure causes <u>air</u> to <u>rush</u> <u>into</u> the balloons — this is like <u>breathing in</u>.

4) Let go of the rubber sheet — this is like <u>relaxing</u> your diaphragm.

5) The <u>volume</u> in the jar gets <u>smaller</u>. This <u>increases</u> the <u>pressure</u>, so air <u>rushes out</u>. Cool, innit...

Left diagram labels: Air rushes in; Balloons fill up like lungs; Pull down

Right diagram labels: Air rushes out; Balloons deflate; relax back up

Inhaling and Exhaling is Breathing In and Out

1) The <u>chest cavity</u> is like a bell jar.

2) When you breathe in, the <u>diaphragm</u> moves <u>down</u> and the <u>ribs</u> move <u>up</u>. This <u>increases</u> the <u>volume</u> of the chest cavity, which <u>decreases</u> the <u>pressure</u>. So air <u>rushes in</u> to fill the lungs.

3) When the diaphragm <u>moves up</u> and the <u>ribs</u> move <u>down</u>, air <u>rushes out</u>.

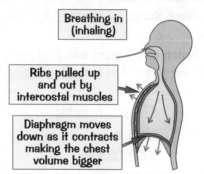

Labels: Breathing in (inhaling); Ribs pulled up and out by intercostal muscles; Diaphragm moves down as it contracts making the chest volume bigger

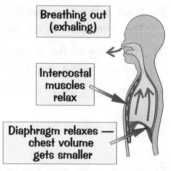

Labels: Breathing out (exhaling); Intercostal muscles relax; Diaphragm relaxes — chest volume gets smaller

Lung Volume Can Be Measured

1) Lung volume is the <u>amount of air</u> you can breathe into your lungs in a single breath.

2) Lung volume is <u>different for different people</u>. For example, <u>taller</u> people tend to have a <u>bigger</u> lung volume than <u>shorter</u> people. And some <u>diseases</u> may <u>reduce</u> a person's lung volume.

3) Lung volume can be <u>measured</u> using a <u>machine</u> called a <u>spirometer</u>.

4) To use a spirometer, a person <u>breathes into the machine</u> (through a tube) for a few minutes. The volume of air that is breathed in and out is measured and plotted on a graph (called a <u>spirogram</u>) like this one...

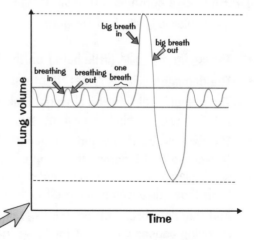

Graph labels: Lung volume (y-axis); Time (x-axis); breathing in; breathing out; one breath; big breath in; big breath out

Now take a deep breath and learn these facts...

Well, if ever you wanted to know how you breathe in and out, now you know. Learn <u>how breathing works</u> — use that <u>bell jar demo</u> to help you understand what goes on in your actual lungs. Make sure you know how <u>lung volume</u> can be measured too. You'll be acing all things breathing in no time at all.

Section 2 — Humans as Organisms

Exercise, Asthma and Smoking

Exercise, asthma and smoking can all affect your gas exchange system and the way in which you breathe.

Exercise

1) When you exercise, your muscles need more oxygen and glucose so they can respire and release energy (see page 4) to keep you going.

2) During exercise, your breathing rate and depth of breathing increase so you can get more oxygen into your blood.

3) If you exercise regularly, the muscles that you use to breathe (the diaphragm and intercostal muscles) will get stronger.

4) This means that your chest cavity can open up more when you breathe in, so you can get more air into your lungs.

5) Over time, regular exercise can also cause an increase in the number and size of the small blood vessels in your lungs and in the number of alveoli. This means gas exchange becomes more efficient.

Asthma

1) People with asthma (asthmatics) have lungs that are too sensitive to certain things (e.g. pet hair, pollen, dust, smoke...).

2) If an asthmatic breathes these things in, the muscles around their bronchioles contract. This narrows the airways.

3) The lining of the airways becomes inflamed and fluid builds up in the airways, making it hard to breathe (an asthma attack).

4) Symptoms of an attack are:
   - difficulty breathing,
   - wheezing,
   - a tight chest.

5) When symptoms appear, sufferers can use an inhaler containing drugs that open up the airways.

Smoking

1) Cigarette smoke contains four main things: carbon monoxide, nicotine, tar and particulates.

2) Tar in particular is really bad for you:

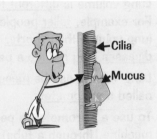
Cilia
Mucus

- Tar covers the cilia (little hairs) on the lining of the airways.
- The damaged cilia can't get rid of mucus properly.
- The mucus sticks to the airways, making you cough more — this is known as smoker's cough.
- The damage builds up and can eventually lead to bronchitis (a disease that inflames the lining of the bronchi) and emphysema (a disease that destroys the air sacs in the lungs). Both these diseases make it difficult to breathe.
- And there's more... tar contains carcinogens (substances that can cause cancer). Smoking causes cancer of the lung, throat and mouth.

This page is just breathtaking...

So there you have it — three different things that have an impact on the gas exchange system. Make sure you know them all. Cover up the page and see how much you can write down about each one. If you're struggling, then re-read the page and try again (and again and again...) until you're confident you know it.

Human Reproductive Systems

Like all <u>mammals</u>, humans have different <u>boy bits</u> and <u>girl bits</u> that allow us to <u>reproduce</u>. No giggling now.

The Male Reproductive System

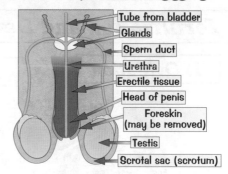

1) <u>Sperm</u> are the male <u>sex cells</u> or '<u>gametes</u>'.
2) Sperm are made in the <u>testes</u> after puberty.
3) Sperm mix with <u>a liquid</u> to make <u>semen</u>, which is <u>ejaculated</u> from the penis during sexual intercourse.

The Female Reproductive System

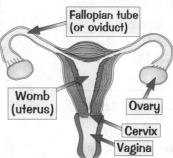

1) An <u>egg</u> is a female <u>sex cell</u> or '<u>gamete</u>'.
2) <u>One</u> of the two ovaries releases an egg <u>every 28 days</u>.
3) It passes into the <u>fallopian tube</u> (or oviduct) where it may <u>meet sperm</u>, which has entered the vagina during <u>sexual intercourse</u> (sometimes known as copulation).
4) If it <u>isn't fertilised</u> by sperm (see next page), the egg will <u>die</u> after about a <u>day</u> and pass out of the vagina.

The Menstrual Cycle Takes 28 Days

1) From the age of puberty females undergo a <u>monthly</u> sequence of events which are collectively known as the <u>MENSTRUAL CYCLE</u>.
2) This involves the body <u>preparing</u> the <u>uterus</u> (womb) in case it receives a <u>fertilised egg</u>.
3) If this doesn't happen, then the egg and uterus lining <u>break down</u> and are <u>lost</u> from the body through the <u>vagina</u> over a period of <u>three</u> to <u>four</u> days, usually.
4) The cycle has <u>four</u> main stages — they are summarised in the diagram and table below:

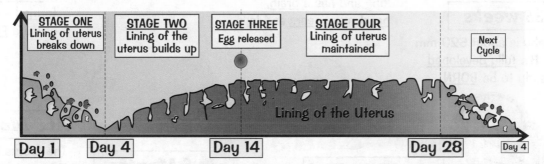

Day	What happens...
1	<u>BLEEDING STARTS</u> as the <u>lining of the uterus</u> (the womb) <u>breaks down</u> and passes out of the vagina — this is what's known as "having a <u>PERIOD</u>".
4	The <u>lining</u> of the uterus starts to <u>build up</u> again. It thickens into a spongy layer full of <u>blood vessels</u> ready for <u>IMPLANTATION</u>. (See next page.)
14	An <u>egg is released</u> from the ovaries of the female, so this is the <u>MOST LIKELY</u> time in which a female may become <u>pregnant</u>. (This day may vary from one woman to the next.)
28	The wall remains thick awaiting the <u>arrival</u> of a <u>fertilised egg</u>. If this doesn't happen then this lining <u>breaks down</u>, passing out of the vagina. Then the whole cycle <u>starts again</u>.

Menstruation — nothing to do with 'men' whatsoever...

There are quite a few details to <u>learn</u> here. <u>Make sure</u> you <u>know</u> all the bits and bobs in the <u>male</u> and <u>female</u> <u>reproductive systems</u>, plus exactly what all the four stages of the <u>menstrual cycle</u> are and when they occur.

Having a Baby

Once Dad's sperm has fertilised Mum's egg, an embryo forms, the gestation period passes, a baby is born.

Fertilisation and Development

1) Ovulation

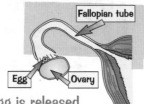

An egg is released from an ovary (around day 14).

2) Copulation

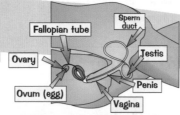

Millions of sperm are released from the penis into the vagina during intercourse.

3) Fertilisation

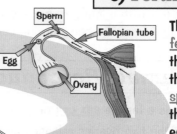

The egg is fertilised when the nuclei of the egg and sperm join — the fertilised egg is called a ZYGOTE.

5) Implantation

About one week after fertilisation, the embryo starts to embed (implant) itself into the wall of the uterus and the placenta begins to develop.

4) Cell Division

24 HOURS after fertilisation the fertilised egg divides into two. After about 4 DAYS the egg has divided into 32 cells. It's now called an EMBRYO.

The Embryo Develops During Gestation

Start here →

At 1 Month

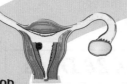

The embryo is 6 mm long and has a brain, heart, eyes, ears and legs.

At 9 Weeks

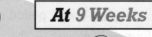

The body is about 25 mm long and is completely formed — it's now called a FOETUS.

At 39 weeks

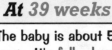

The baby is about 520 mm long. It's fully developed and ready to be BORN.

At 7 Months

The foetus is 370 mm long and is 'VIABLE'. This means it would have a fair chance of surviving if it were born at this stage.

At 5 Months

It's now about 160 mm long. It kicks and its pesky finger nails can be felt.

At 3 Months

The foetus is 54 mm long and looks much more like a baby.

The Mother's Lifestyle During Pregnancy is Important

1) The placenta lets the blood of the foetus and mother get very close to allow exchange of food, oxygen and wastes.

2) If the mother smokes, drinks alcohol or takes other drugs while she is pregnant, harmful chemicals in her blood can cross the placenta and affect the foetus. For example, the foetus may not develop properly and could have health problems after it's born.

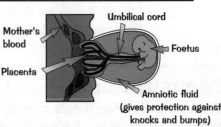

Umbilical cord
Mother's blood
Placenta
Foetus
Amniotic fluid (gives protection against knocks and bumps)

Health and Drugs

Good health is a situation where you're <u>fine and dandy</u> both <u>physically</u> and <u>mentally</u>.
Recreational drugs can have serious <u>negative effects</u> on your health.

Health is More Than Just the Absence of Disease

Good health means having <u>BOTH</u> of these:

1) A <u>healthy body</u> that's <u>all working properly</u> with <u>no diseases</u>.

2) A <u>healthy mental state</u> where you're able to cope with the <u>ups and downs</u> of life.

You should <u>look after your body</u> by eating a <u>balanced diet</u>, doing enough <u>exercise</u> and <u>not abusing drugs</u>.

Drugs

1) A drug is anything that <u>affects the way</u> the body works.
E.g. They may raise the heart rate or cause blurred vision.

2) There are <u>LEGAL DRUGS</u> and <u>ILLEGAL DRUGS</u>.
Aspirin, caffeine and antibiotics are examples of <u>legal drugs</u>.
Cannabis, speed and ecstasy are examples of <u>illegal drugs</u>.

3) <u>RECREATIONAL DRUGS</u> are drugs used for fun.
They can be legal or illegal.

4) Drugs can affect <u>life processes</u>. For example, drugs that affect
the <u>brain</u> are likely to affect <u>movement</u> and <u>sensitivity</u>. And drugs that affect
the <u>liver and kidneys</u> will most likely affect <u>excretion</u> (as these are the organs that process waste).

> **7 Life Processes**
> Movement — moving parts of the body.
> Reproduction — producing offspring.
> Sensitivity — responding and reacting.
> Nutrition — getting food to stay alive.
> Excretion — getting rid of waste.
> Respiration — releasing energy from food.
> Growth — getting to adult size.

Solvents

1) Solvents are found in most homes — in things like <u>paints</u>, <u>aerosols</u> and <u>glues</u>.

2) They're drugs because they cause <u>hallucinations</u>, which are illusions of the mind.
Solvents usually have a severe effect on <u>behaviour</u> and <u>character</u>.

3) They also cause serious <u>damage</u> to the <u>lungs</u>, the <u>brain</u>, <u>liver</u> and <u>kidneys</u>.

Alcohol

1) Alcohol is found in <u>beers</u>, <u>wines</u> and <u>spirits</u>. It's <u>illegal</u> to buy it <u>under the age of 18</u>.

2) It's a <u>depressant</u>, which means it <u>decreases the activity of the brain</u> and <u>slows down responses</u>.

3) It's a <u>poison</u> which affects the <u>brain</u> and <u>liver</u> leading to
various health problems, e.g. <u>cirrhosis</u> (liver disease).

4) It <u>impairs judgement</u> which can lead to <u>accidents</u>. It's also very <u>addictive</u>.

Illegal Drugs — Dangerous, Addictive and Life-Wrecking

1) Ecstasy and LSD are <u>hallucinogens</u>. Ecstasy can give the feeling of <u>boundless
energy</u> which can lead to <u>overheating</u>, <u>dehydration</u> and sometimes <u>DEATH</u>.

2) Heroin and Morphine were developed as <u>painkillers</u>. However they turned out to be
highly <u>addictive</u>. They can both cause severe <u>degeneration</u> of a person's life.

3) Amphetamine (speed) and Methedrine are <u>stimulants</u>. They give a feeling of
<u>boundless energy</u>. However, users quickly become <u>psychologically dependent</u> on
the drug (i.e. they think they <u>need</u> them), so <u>behaviour</u> and <u>character</u> deteriorate.

4) Barbiturates are <u>depressants</u>. They <u>slow down</u> the nervous system and therefore
<u>slow down</u> reaction time. They can help <u>sleeping</u> but they're <u>seriously habit-forming</u>.

Drugs aren't harmless fun — they're a slippery slope...

There are lots of details here. Make sure you know how different <u>recreational drugs</u> can affect <u>behaviour</u>,
<u>health</u> and <u>life processes</u>. And above all else, make sure you realise the <u>dangers</u> of illegal drugs. <u>Learn it well</u>.

Section Summary

Well, there's certainly some stuff in Section 2 — all you ever wanted to know about human beings, and a good deal more besides I should think. Now what you've got to do is make sure you learn it all. And here again for your delectation and enjoyment I have prepared some more of those splendid questions. I didn't say exciting questions — just splendid. Remember, you have to keep coming back to these questions time and time again, to see how many of them you can do. All they do is test the basic simple facts. OK then — let's see how much you've learnt so far...

1) Name all five nutrients in a balanced diet.
2) Say what each nutrient is important for in the body.
3) For each of the five nutrients, give three examples of foods that contain them.
4) Apart from the five nutrients, give two things that are needed in a balanced diet and explain why they're needed.
5) Give two things that affect how much energy a person needs each day.
6)* Sonia has a body mass of 54 kg. What is her daily basic energy requirement?
7) What is obesity? How is it caused?
8) What health problems can be caused by getting too little food?
9) What does digestion do?
10) Name eight main bits of the alimentary canal.
11) Say what goes on in each of the eight bits.
12) Why can't big molecules pass through gut walls? What has to happen to them first?
13) What are villi? What is their function (job) and how are they well-suited to it?
14) Give four reasons why the bacteria found naturally in your digestive system are good news.
15) Give four functions of the skeleton.
16) What are antagonistic muscles?
17) Explain in terms of "muscle contraction" how you can move your arm up and down.
18) What is a moment? What two pieces of information do you need to be able to calculate one?
19)* Calculate the force applied by the muscle to keep the arm still in this diagram.
20) Sketch the human gas exchange system and label all the important structures.
21) What gases are exchanged in the lungs when air is breathed in? Where does each gas move from and to?
22) Give three ways in which the lungs are well-adapted for gas exchange.
23) Explain how we breathe air in and out.
24) How can lung volume be measured?
25) How does exercise affect the gas exchange system?
26) What happens in the gas exchange system when someone has an asthma attack?
27) What are the symptoms of an asthma attack?
28) Give two ways in which smoking affects the gas exchange system.
29) What are the male sex cells called? Where are they made?
30) And what are the female sex cells called? Where are they made?
31) Outline the four main stages of the menstrual cycle and say when they happen.
32) What exactly is fertilisation?
33) List five things which must happen before a human embryo can start to develop.
34) Describe what an embryo looks like at:
   1 month, 9 weeks, 3 months, 5 months, 7 months, 39 weeks.
35) Explain why it's not a good idea for a woman to smoke while she's pregnant.
36) What does being healthy mean?
37) What are drugs? What is meant by a 'recreational' drug?
38) Name three legal drugs and three illegal drugs.   (Steer clear pal.)
39) Name one recreational drug and explain how it affects life processes.

*Answers on page 108.

Plant Nutrition

THINK about this: plants make their own food — it's a nice trick if you can do it.

Photosynthesis *Makes* Food *From* Sunlight

1) Photosynthesis is a chemical process which takes place in every green plant.
2) Photosynthesis basically produces food — in the form of glucose (a carbohydrate).
3) The plant can then use the glucose to increase its biomass — i.e. to grow.
4) Photosynthesis happens in all the green bits of a plant but mainly in the leaves.

Four Things *are* Needed *for Photosynthesis...*

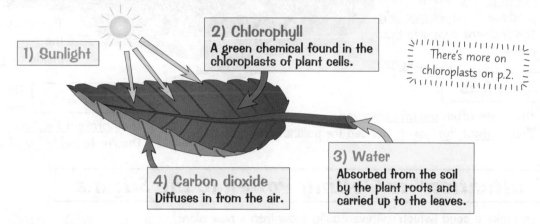

1) Sunlight

2) Chlorophyll
A green chemical found in the chloroplasts of plant cells.

There's more on chloroplasts on p.2.

3) Water
Absorbed from the soil by the plant roots and carried up to the leaves.

4) Carbon dioxide
Diffuses in from the air.

Chlorophyll absorbs sunlight and uses the energy to convert carbon dioxide and water into glucose. Oxygen is also produced. This word equation summarises what happens during photosynthesis. Learn it:

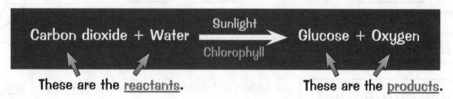

$$\text{Carbon dioxide} + \text{Water} \xrightarrow[\text{Chlorophyll}]{\text{Sunlight}} \text{Glucose} + \text{Oxygen}$$

These are the reactants.   These are the products.

Leaves *are* Adapted *for Efficient Photosynthesis*

Leaves are really good at carrying out photosynthesis. Here's why...

1) Leaves are broad, so there's a big surface area for absorbing light.
2) Most of the chloroplasts are found in cells near the top of the leaf, where they can get the most light.
3) The underside of the leaf is covered in tiny holes called stomata. These holes allow carbon dioxide to diffuse (move) into the leaf from the air. They also allow oxygen to diffuse out. Air spaces inside the leaf allow carbon dioxide to move easily between the leaf cells.
4) Leaves also contain a network of veins, which deliver water to the leaf cells and take away glucose.

Plants Also Need Minerals *from the Soil*

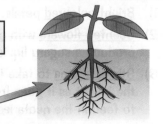

1) Plants grow using the food they make themselves in photosynthesis. But to keep healthy they also need mineral nutrients from the soil.
2) Plants absorb these minerals through their roots (along with water).

Hmm, it's all clever stuff — just make sure you learn it...

Remember, plants don't get food from the soil — they make it themselves using photosynthesis. They do get water and tiny amounts of the minerals they need to stay healthy from the soil. Got that? Sorted.

Plant Reproduction

Just like humans, <u>plants reproduce</u> (<u>make babies</u>). This page is all about how plant reproduction starts.

The Flower Contains the Reproductive Organs

1) Stamens

The sta-<u>men</u>-s are the <u>male</u> parts of the flower. They consist of the <u>anther</u> and the <u>filament</u>. The anther contains <u>pollen grains</u>, which produce the <u>male sex cells</u>. The filament supports the anther.

2) Carpels

The <u>female</u> parts of the flower. They consist of the

<u>stigma</u>,

<u>style</u> and

<u>ovary</u>.

The ovary contains the <u>female sex cells</u> inside <u>ovules</u>.

3) Petals

These are often <u>brightly coloured</u>. They <u>attract the insects</u> needed for pollination.

4) Sepals

These are <u>green</u> and <u>leaf-like</u>. They <u>protect the flower</u> in the <u>bud</u>. They're found <u>below</u> the main petals.

"Pollination" is Getting Pollen to the Stigma

1) To make a <u>seed</u> (which will eventually grow into a new plant) the <u>male</u> and <u>female sex cells</u> must "<u>meet up</u>".

2) To do this, the <u>pollen grains</u> must get from a <u>stamen</u> to a <u>stigma</u>. This can happen in <u>two ways</u>:

1) Self Pollination

— pollen is transferred from <u>stamen</u> <u>to stigma</u> on the <u>SAME PLANT</u>.

2) Cross Pollination

— pollen is transferred from the <u>stamen</u> of <u>one plant</u> to the <u>stigma</u> of a <u>DIFFERENT PLANT</u>. Cross pollination can involve...

...Insect Pollination

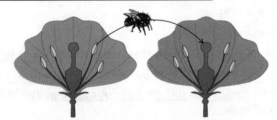

<u>Plant features</u> that help <u>insect pollination</u>:

1) <u>Bright coloured</u> petals.
2) <u>Scented flowers</u> with <u>nectaries</u> (glands that produce a sugary liquid for insects to feed on).
3) <u>Sticky stigma</u> to take the pollen off the insect as it goes from plant to plant to feed in the nectaries.

...Wind Pollination

<u>Features</u> of plants that use <u>wind pollination</u>:

1) <u>Usually small dull petals</u> on the flower.
2) <u>No scent</u> or nectaries.
3) <u>Long filaments</u> hang the anthers outside the flower so a lot of <u>pollen</u> is <u>blown away</u>.
4) Stigmas are <u>feathery</u> to <u>catch pollen</u> as it's carried past in the wind.

There are no "B"s in wind pollination — bzzzz bzzz...

<u>All those bits</u> of a <u>flower</u> to learn and then the "ins and outs" of <u>pollination</u>. Make sure you <u>know it all</u> before you move on to the next page or you'll be in a pretty <u>pickle</u>. And not the edible kind either...

Fertilisation and Seed Formation

Here it is, the long awaited <u>sequel</u> to <u>Plant Reproduction</u> — or what happens <u>after</u> a flower is <u>pollinated</u>.

Fertilisation *is the* Joining *of Sex Cells*

1) <u>Pollen</u> is the <u>plant equivalent</u> of human <u>sperm</u>.

2) <u>Pollen grains</u> land on a <u>ripe stigma</u> with help from insects or the wind (see previous page).

3) A <u>pollen tube</u> then grows out of a pollen grain, down through the <u>style</u> to the <u>ovary</u>.

4) The <u>nucleus</u> from a male sex cell <u>moves down</u> the tube to <u>join</u> with a female sex cell inside an <u>ovule</u>. <u>Fertilisation</u> is when the <u>two nuclei join</u>.

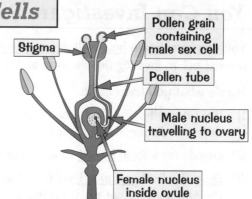

Stigma

Pollen grain containing male sex cell

Pollen tube

Male nucleus travelling to ovary

Female nucleus inside ovule

Seeds *are* Formed *From Ovules*

1) After fertilisation, the <u>ovule</u> develops into a <u>seed</u>. Each seed contains a <u>dormant</u> (inactive) <u>embryo plant</u>.

2) The embryo has a <u>food store</u> which it uses when conditions are right — i.e. it starts to <u>grow</u> or "<u>germinate</u>".

3) The <u>ovary</u> develops into a <u>fruit</u> around the seed. Fruits can tempt animals to <u>eat them</u> and so <u>scatter the seeds</u> in their faeces ("poo").

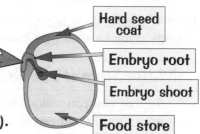

Hard seed coat

Embryo root

Embryo shoot

Food store

Seed Dispersal *is Scattering* Seeds

Seeds are <u>dispersed</u> or <u>spread out</u> so that they can grow <u>without</u> too much <u>competition</u> from <u>each other</u>. Here are some ways in which the seed can be dispersed:

1) Wind *dispersal*

<u>Dandelion</u> fruit.

<u>Parachutes</u> catch the wind.

<u>Sycamore</u> fruit.

<u>Wings</u> help it fly away from the parent tree.

2) Animal *dispersal*

<u>Tomato</u> fruit.

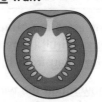

Fruit gets <u>eaten</u>. Seeds come out in the animals' <u>droppings</u>.

<u>Burdock</u> fruit.

<u>Hooks</u> catch animals' coats.

3) Explosions

Peas.

The pods <u>dry out</u> and <u>flick</u> the seed out.

4) Drop and Roll

The heavy fruit <u>falls</u> down from the tree. It <u>splits</u> when it hits the ground and the seeds <u>roll</u> out.

<u>Horse Chestnut</u> fruit.

The seeds then tend to be further dispersed by animals.

What has a hazelnut in every bite — Squirrel poo...

Phew, what a palaver. It all starts with <u>pretty flowers</u> which lead to <u>pollination</u>, then <u>fertilisation</u>. That leads on to <u>seed development</u>, and then the business of <u>dispersal</u>. Eventually, the seeds will <u>grow</u> into <u>new plants</u> far away from their parents. It makes your head hurt just thinking about it. You do need to <u>learn it</u> though.

Investigating Seed Dispersal Mechanisms

At last, a little bit of science in action. Roll up your sleeves and let's get started.

You Can Investigate Seed Dispersal by Dropping Fruit

You can investigate how well different seeds disperse from the comfort of your own classroom.
It's easiest to investigate the wind and drop and roll dispersal mechanisms.

Here's what you have to do.

1) Get yourself some fruit (containing seeds). You could compare ones with different dispersal mechanisms, e.g. sycamore fruit and horse chestnut fruit.

2) Decide on a fixed height to drop the fruit from.

3) Drop the fruit one at a time from this height, directly above a set point on the ground.

4) Using a tape measure, measure and record how far along the ground the seeds have been dispersed.

Seed Type	Distance Dispersed (cm)		
Sycamore	20	25	
Horse Chestnut			

Do this at least three times for each type of seed and then find the average distance each type travels or 'disperses' when dropped.

Make Sure it's a Fair Test

So that you can make a fair comparison between the distances travelled by different seed types, you need to keep the following the same each time you do the experiment:

* the person dropping the fruit,
* the height the fruit are dropped from,
* the place you're doing the experiment (stay away from doors and windows that might cause draughts).

This is called "controlling the variables".

Use a Fan to Investigate the "Wind Factor"

Many seed dispersal mechanisms are affected by the wind.
The special shape of sycamore fruit helps the wind to catch the fruit and carry the seeds far away from the parent sycamore tree.

I want you out from under my feet kids.

You can investigate just how much the wind affects seed dispersal by introducing an electric fan into the experiment above.
Here's how:

1) Set up the fan a fixed distance from the person dropping the fruit.

2) Switch the fan on — it needs to be set to the same speed for every fruit you drop. This makes sure the experiment will be a fair test.

3) Drop the fruit as before and measure how far along the ground the seeds travel.

You should find that the sycamore seeds travel much further in windy conditions (i.e. when the fan is switched on).
This might not be the case for every seed type though.

I've got a pea shooter — is that a seed dispersal mechanism?

Ahh, the things you get up to in science lessons. You can also investigate how much the shape of sycamore fruit helps the seeds to disperse — just cut the wings off and repeat the experiment in front of the fan.

Dependence on Other Organisms

It probably won't surprise you to learn that organisms <u>depend</u> on <u>other organisms</u> for their <u>survival</u>.

Organisms *in an* Ecosystem *are* Interdependent

1) An <u>ecosystem</u> is <u>all</u> the <u>living organisms</u> in <u>one area</u>, plus their <u>environment</u>.
2) The <u>organisms</u> in an ecosystem are <u>interdependent</u> — they <u>need each other</u> to survive.

Almost All Living Things Depend *on* Plants

Almost <u>all life on Earth</u> depends on <u>plants</u>. Without them, we just wouldn't be here. Here's why...

Plants Capture *the* Sun's Energy

1) <u>Almost all energy</u> on <u>Earth</u> comes from the <u>Sun</u>.
2) <u>Plants</u> use some of the Sun's energy to <u>make food</u> during <u>photosynthesis</u> (see page 19). They then use this food to build "<u>organic molecules</u>" (things like carbohydrates and proteins), which become part of the plants' cells.
3) These organic molecules <u>store</u> the Sun's energy. The energy gets <u>passed on from plants to animals</u> when animals <u>eat</u> the plants. It gets <u>passed on again</u> when these animals are <u>eaten</u> by <u>other animals</u>.
4) <u>Only plants</u>, <u>algae</u> (seaweeds) and some <u>bacteria</u> are <u>able</u> to carry out <u>photosynthesis</u>. So nearly all living things <u>rely on plants</u> to <u>capture</u> and <u>store</u> the <u>Sun's energy</u>.

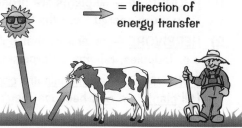

= direction of energy transfer

Plants Release Oxygen *and* Take in Carbon Dioxide

1) All living things <u>respire</u> (see page 4).
2) When plants and animals respire, they <u>take in oxygen</u> (O_2) from the atmosphere and <u>release carbon dioxide</u> (CO_2).
3) When plants <u>photosynthesise</u>, they do the <u>opposite</u> — they <u>release oxygen</u> and <u>take in carbon dioxide</u>.
4) So photosynthesis helps make sure there's always <u>plenty of oxygen</u> around for respiration. It also helps to <u>stop</u> the <u>carbon dioxide level</u> in the atmosphere from getting <u>too high</u>. This is an example of <u>organisms affecting</u> their <u>environment</u>.

Many Plants Depend *on* Insects *in Order to* Reproduce

1) Many plants depend on insects to <u>pollinate</u> them (see page 20).
2) Without insects like <u>bees</u>, <u>moths</u> and <u>butterflies</u>, these plants would <u>struggle</u> to <u>reproduce</u>.
3) This would obviously be <u>bad</u> for the <u>plants</u>, but it would be <u>bad</u> for <u>humans</u> too. Many of our <u>crop plants</u> need to be pollinated by insects in order to produce the <u>fruit</u>, <u>nuts</u> and <u>seeds</u> that <u>we eat</u>.
4) So we depend on insects to pollinate our crops and <u>ensure</u> our <u>food supply</u>.

We're all just one big happy family...

Remember, the organisms in an ecosystem are <u>interdependent</u> — we <u>depend</u> on <u>plants</u> for all our <u>energy</u> and to produce the <u>oxygen</u> we use up in respiration. And many plants depend on <u>insects</u> to <u>pollinate</u> them.

Food Chains and Food Webs

Organisms depend on each other to <u>survive</u>. Mainly this means that they depend on each other for <u>food</u>.

Food Chains Show What is Eaten by What

1) The organisms in a food chain are usually in the <u>same ecosystem</u>.

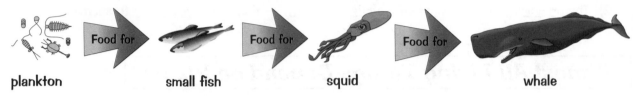

plankton   small fish   squid   whale

2) The <u>arrows</u> show what is eaten by what — i.e. "<u>food for</u>". (Plankton is <u>food for</u> small fish, etc.)

3) The arrows also show the <u>direction of energy flow</u>.

Food Webs and Their Tremendous Terminology

Food webs contain <u>many</u> interlinked <u>food chains</u> as shown here:
Learn these <u>nine bits</u> of <u>terminology</u>:

1) <u>PRODUCER</u> — all <u>plants</u> are <u>producers</u>. They store energy from the Sun.

2) <u>HERBIVORE</u> — an animal that <u>only eats plants</u>, e.g. tadpoles, rabbits, caterpillars, aphids.

3) <u>CONSUMER</u> — all <u>animals</u> are <u>consumers</u>. (All <u>plants</u> are <u>not</u>, because they're producers.)

4) <u>PRIMARY CONSUMER</u> — an animal that eats <u>producers</u> (plants).

5) <u>SECONDARY CONSUMER</u> — an animal that eats primary consumers.

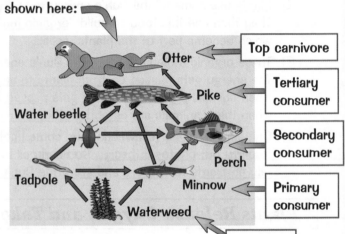

6) <u>TERTIARY CONSUMER</u> — an animal that eats secondary consumers.

7) <u>CARNIVORE</u> — eats <u>only animals</u>, never plants.

8) <u>TOP CARNIVORE</u> — is <u>not eaten by anything else</u>.

9) <u>OMNIVORE</u> — eats <u>both plants and animals</u>.

The organisms in a food web are all <u>interdependent</u> — so a <u>change</u> in <u>one organism</u> can easily <u>affect others</u>.

<u>Example</u> — What happens if the minnows are removed?

1) Who <u>will get eaten LESS</u>? The <u>tadpoles</u>, as there are no minnows there to eat them.
2) Who <u>will get eaten MORE</u>?   a) <u>Water beetles</u> (by perch who'll get hungry without minnows).
           b) <u>Waterweeds</u> (since the numbers of tadpoles will increase).

Poisons Build Up as They are Passed Along a Food Chain

= level of poison

<u>Toxic materials</u> (poisons) can sometimes get into food chains and <u>harm</u> the organisms involved. Organisms <u>higher up</u> the food chain (usually the <u>top carnivores</u>) are likely to be the <u>worst affected</u> as the <u>toxins accumulate</u> (build up) as they are passed along.

Learn about Food Webs — but don't get tangled up...

Once you've learnt everything on this page, you can practise this <u>typical food web question</u>: "If the number of otters decreased, give one reason why the number of water beetles might a) decrease   b) increase".*

Section Summary

Green plants are ace aren't they? What I really like about them is that they're all so clean and fresh — human and animal biology always seems to end up so gory with all sorts of gruesome diagrams and horrid diseases. But plants have such simple lives — they just seem to "go with the flow", with no apparent discomfort and no worries — and let's face it, it's a nice trick if you can do it.

Alas nature conspired to give humans an altogether more "challenging" experience on this little blue-green planet of ours — and somehow that's ended up with you needing to know the answers to all these questions. Hmmm, it's a funny old world isn't it — when you think about it from that angle...
Anyway, here they are. Off you go then...

1) What is made during photosynthesis?

2) What do plants do with glucose?

3) What are the four things needed for photosynthesis to happen?

4) Write out the word equation for photosynthesis.

5) What is the by-product made in photosynthesis, which is needed by animals?

6) Apart from water, what do plants need from the soil?

7) What are the four main parts of a flower? Say what each part actually does.

8) What is pollination? What are the two types of pollination?

9) What is the difference between insect pollination and wind pollination?

10) Give three features of: a) an insect pollinated plant,   b) a wind pollinated plant.

11) What is fertilisation? How does the male sex cell nucleus get from the stigma to the ovary?

12) What does an ovule develop into after fertilisation?

13) What does the ovary eventually develop into?

14) Give another name for seed dispersal.

15) Give four ways in which seeds can be dispersed.
    Give an example of a fruit that disperses seeds in each of these ways.

16) Describe how you could investigate the seed dispersal mechanism of a sycamore tree.
    How could you investigate the effect of wind on this dispersal mechanism?

17) What is an ecosystem?

18) Explain why most living things rely on plants for energy.

19) What else do living things rely on plants for?

20) What do many plants rely on insects for? How does this affect us humans?

21) What is a food chain? And a food web?

22) Give good definitions for all of the following terms:
    a) producer   b) herbivore   c) consumer
    d) primary consumer   e) secondary consumer   f) tertiary consumer
    g) carnivore   h) top carnivore   i) omnivore.

23) What happens to poisons as they are passed along a food chain?

DNA and Inheritance

DNA's brilliant stuff — it's like your body's own instruction manual. When you're being made, you get bits of DNA from your mum and bits from your dad — this is how you inherit characteristics.

Chromosomes, DNA and Genes

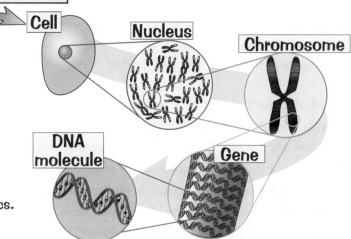

1) Most cells in your body have a nucleus. The nucleus contains chromosomes.

2) Chromosomes are long, coiled up lengths of a molecule called DNA.

3) DNA is a long list of chemical instructions on how to build an organism.

4) A gene is a short section of a chromosome (and so a short section of DNA).

5) Genes control many of our characteristics, e.g. hair colour, eye colour, hairiness, etc. Different genes control different characteristics.

6) Genes work in pairs — one will usually be dominant over the other.

We Inherit Characteristics From Our Parents

1) Human body cells have 46 chromosomes (23 pairs).

2) Sperm and egg cells carry only 23 chromosomes.

3) During reproduction, when an egg is fertilised, the nucleus of the egg fuses with the nucleus of the sperm.

4) This means that the fertilised egg contains 23 matched pairs of chromosomes. It has one copy of each gene from the mother and one from the father.

5) Since genes control characteristics, the fertilised egg develops into an embryo with a mixture of the parents' characteristics. This is how you 'inherit' your parents' characteristics.

6) The process by which genes are passed down from parents to their offspring is called heredity.

A characteristic passed on in this way is called a 'hereditary' characteristic.

The Structure of DNA Was Only Worked Out Recently

1) Scientists struggled for decades to work out the structure of DNA.

2) Crick and Watson were the first scientists to build a model of DNA — they did it in 1953.

3) They used data from other scientists, Wilkins and Franklin, to help them understand the structure of the molecule. This included X-ray data showing that DNA is a double helix — a spiral made of two chains wound together.

One chain

Another chain

4) By putting all the information together, Crick and Watson were able to build a model showing what DNA looks like.

DNA? I get all my genes from Topshop...

I can see three main headings, sixteen numbered points and two important diagrams — and they all need learning. Sitting down and challenging yourself to repeat the main details from this page really isn't that hard. If you're struggling with any bits you can re-read the page, then cover it back up and try again.

Variation

This page is all about <u>differences between organisms</u> — both <u>big, obvious differences</u>, like those between a tree and a cow, and <u>less obvious differences</u>, like people having different blood groups.

Different *Species* Have Different *Genes*

1) All living things in the world are <u>different</u> — we say that they show <u>VARIATION</u>.

2) A human, a cow, a dandelion and a tree all look different because they're different <u>species</u>. These differences <u>between</u> species occur because their <u>genes</u> are very <u>different</u>.

3) But you also see variation <u>within a species</u>, i.e. plants or animals that have <u>basically</u> the <u>same</u> genes will also show differences between them, e.g. skin colour, height, flower size, etc. Any <u>difference</u> is known as a <u>characteristic feature</u>.

4) Characteristic features can be <u>inherited</u> (come from your parents via genes) or they can be <u>environmental</u> (caused by your surroundings).

Continuous **and** *Discontinuous* **Variation**

Variation <u>within a species</u> can either be classed as <u>continuous</u> or <u>discontinuous</u>.

Continuous Variation — *the feature can vary over a* range of values

1) Examples of this are things like <u>height</u>, <u>weight</u>, <u>skin colour</u>, <u>intelligence</u>, <u>leaf area</u>, etc. where the feature can have <u>any value at all</u> — within a certain <u>range</u>. If you did a survey of kids' heights you could plot the results on a chart like the one opposite (the heights would be collected into groups to give the bars).

2) The smooth <u>distribution curve</u> drawn on afterwards (the red line) shows much better the <u>continuous</u> way that values for height actually vary.

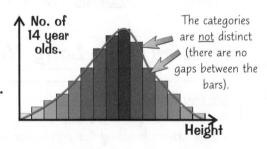

The categories are <u>not</u> distinct (there are no gaps between the bars).

Discontinuous Variation — *the feature can only take* certain values

1) An example of this is a person's <u>blood group</u>, where there are just <u>four distinct options</u>, <u>NOT</u> a whole <u>continuous range</u>.

2) Another example is the <u>colour of a courgette</u>. A courgette is either yellow, light green or dark green — there's <u>no range</u> of values.

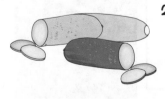

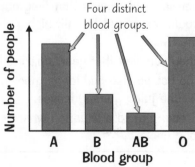

Four distinct blood groups.

Variationvariationvariation — *guess what type of variation that is...*

Don't let the fancy word "variation" put you off. People seem to think it must mean something really complicated. It doesn't. It just means "<u>differences</u>" (between any living things). You can have variation (differences) <u>between</u> different species, and you can also have variation (differences) <u>within</u> one species.

Natural Selection and Survival

It's OK to be different — that's what our parents are always trying to tell us. In nature, being different can be really important. Having a different characteristic to other organisms can determine whether an organism (and its future generations) are likely to survive in the long run or not. It's serious stuff.

Variation Leads to Natural Selection

1) Organisms show variation because of differences in their genes (see previous page).

2) Organisms also have to compete for the resources they need in order to survive and reproduce, e.g. food, water and shelter. They have to compete with other members of their own species, as well as organisms from other species.

FOR EXAMPLE...
...this red squirrel...

...has to compete with other red squirrels (members of its own species)...

...as well as grey squirrels (a different species), in order to get food.

3) Organisms with characteristics that make them better at competing are more likely to survive and reproduce. This means they're more likely to pass on the genes for their useful characteristics to the next generation.

4) Organisms that are less successful competitors are usually the first to die — possibly before they've had a chance to reproduce. This means their genes and less useful characteristics won't be passed on to any offspring.

5) So, over time, the gene for a useful characteristic will become more common.

6) This process in which a characteristic gradually becomes more (or less) common in a population is known as natural selection.

A population is all the organisms of one species that live in the same ecosystem.

Giraffes Have Long Necks Due to Natural Selection

Are you sitting comfortably...

Once upon a time there was a group of animals munching leaves from a tree. Unfortunately the population was high and food was running short.

Soon all the leaves on the lower parts of the trees were gone and the animals started to get hungry — some even died. Except, that is, for a couple of animals which happened to have slightly longer necks than normal. This meant that they could compete better for food — they could reach just that bit higher, to the juicy and yummy leaves higher up the trees.

They survived that year, unlike a lot of animals, and had lots of babies. The babies also had longer necks, and could eventually reach up the tree for the juicy yummy leaves.

It soon got to a situation where most of the animals in the population had long necks...

Survival of the Giraffe — a tall story...

Make sure you get the gist of this. Only those who are born with features that make them great at competing in the world they live in will survive and produce offspring — the sick and the inept all die off very quickly. Just be glad you're a human and live in the time and the place that you do!

Extinction and Preserving Species

Organisms that can't compete <u>don't survive</u> for long. It's a <u>cruel world</u> out there.

Many Species *Are at Risk of Becoming* Extinct

1) Many organisms <u>survive</u> because they are <u>well-adapted</u> for <u>competing</u> in their environment.

2) But if the environment <u>changes</u> in some way, some organisms may struggle to <u>compete successfully</u> for the resources they need to <u>survive</u> and <u>reproduce</u>.

3) If this happens to a <u>whole species</u>, then that species is at risk of becoming <u>extinct</u>. <u>Extinct</u> means that there are <u>none of them left at all</u> (like the woolly mammoth).

4) Species <u>at risk</u> of becoming extinct are called <u>endangered species</u>.

Humans *Can* Suffer *When Species Become* Extinct

1) Humans <u>rely</u> on <u>plants</u> and <u>animals</u> for <u>food</u>.

2) We also use them to make <u>clothing</u>, <u>medicines</u>, <u>fuel</u>, etc.

3) We need to <u>protect</u> the organisms we already use in this way. We also need to make sure organisms we <u>haven't discovered yet</u> don't become extinct before we find them — or we might <u>miss out</u> on <u>new sources</u> of useful products.

4) <u>Ecosystems</u> are <u>complex</u>. If <u>one species</u> becomes <u>extinct</u>, this can have a <u>knock-on effect</u> for <u>other organisms</u> — including <u>us</u>.

5) That's why it's important for us to <u>maintain</u> the planet's <u>biodiversity</u> — the <u>variety</u> of <u>species</u> that live on Earth.

There are probably loads of species we don't know about, e.g. in unexplored rainforests and deep in the ocean.

Gene Banks *May Help to* Prevent Extinction

1) A <u>gene bank</u> is basically a <u>store</u> of the <u>genes</u> of different species.

2) This means that if a species becomes <u>endangered</u> or even <u>extinct</u>, it may be possible to <u>create new members</u> of that species. So gene banks could be a way of <u>maintaining biodiversity</u> in the future.

3) Genes are stored differently for plants and animals. For example:

PLANTS	ANIMALS
<u>Seeds</u> (which contain genes) can be <u>collected</u> from plants and <u>stored</u> in <u>seed banks</u>.	<u>Sperm</u> and <u>egg cells</u> (which contain genes) may be <u>frozen</u> and <u>stored</u>.
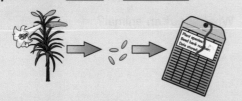	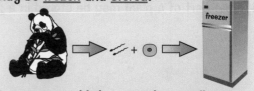
If the plants become <u>extinct</u> in the wild, <u>new plants</u> can be <u>grown</u> from the seeds kept in storage.	Scientists could then use these cells to create new <u>animal embryos</u> in the future.

Gene banks <u>aren't</u> the <u>only way</u> to maintain biodiversity. It's much <u>better</u> to try to <u>stop</u> species becoming <u>extinct</u> in the <u>first place</u>, e.g. by <u>preventing</u> the <u>destruction of habitats</u> (the areas where organisms live).

My bank balance is two seeds and an egg...

You need to understand how <u>environmental changes</u> can put species at risk of <u>extinction</u>. Also, make sure you understand why it's so important to <u>maintain biodiversity</u> and how <u>gene banks</u> may help to do this.

Section Summary

Section 4 is fairly basic stuff really, but there are one or two fancy words which might cause you quite a bit of grief until you've made the effort to learn exactly what they mean: "DNA" is just a list of instructions for how any living creature is put together; "variation" just means "differences", etc., etc.

These questions aren't the easiest you could find, but they test exactly what you know and find out exactly what you don't. You need to be able to answer them all, because all they do is test the basic facts.

You must practise these questions over and over again until you can just sail through them.

1) Where do you find chromosomes?
2) What are chromosomes made of?
3) What is a gene? What do genes control?
4) How many chromosomes do humans have in each body cell?
5) How many chromosomes are there in human sperm cells? How about in human egg cells?
6) What happens at fertilisation?
7) What does heredity mean?
8) Name the two scientists who first built a model of DNA.
   Name the other two scientists whose data helped them.
9) Describe the structure of a DNA molecule.
10) What does variation mean?
11) Why do different species look different?
12) What is a characteristic feature?
13) What is continuous variation? Give three examples.
14) What is discontinuous variation? Give two examples.
15) Give one way in which a graph showing continuous variation would
    differ from a graph showing discontinuous variation.
16) Why is it important that organisms are good at competing for the things they need?
17) Why are genes for useful characteristics likely to become more common in a population over time?
    What is this process called?
18) How did giraffes end up with very long necks?
19) Why could it be bad news for an organism if its environment changes?
20) What does extinct mean?
21) What does endangered mean?
22) What is biodiversity? Why is it important for us to maintain the planet's biodiversity?
23) What is a gene bank? What are they used for?
24) What part of a plant may be stored in a gene bank? What about an animal?

Solids, Liquids and Gases

The first page in this section is all about <u>states of matter</u> and there are only <u>three</u> you need to know.

The Three States of Matter — Solid, Liquid and Gas

1) Materials come in <u>three</u> different forms — <u>solids</u>, <u>liquids</u> and <u>gases</u>.
2) These are called the <u>Three States of Matter</u>.
3) All <u>materials</u> are made up of <u>tiny particles</u>.
4) Which <u>state</u> you get (<u>solid</u>, <u>liquid</u> or <u>gas</u>) depends on how <u>strongly</u> the particles <u>stick together</u>. How well they stick together depends on <u>three things</u>:
      a) the <u>material</u>   b) the <u>temperature</u>   c) the <u>pressure</u>.

Solids, Liquids and Gases Have Different Properties

1) We can <u>recognise</u> solids, liquids and gases by their different <u>properties</u>.
2) A <u>property</u> of a substance is just a way of saying <u>how it behaves</u>.

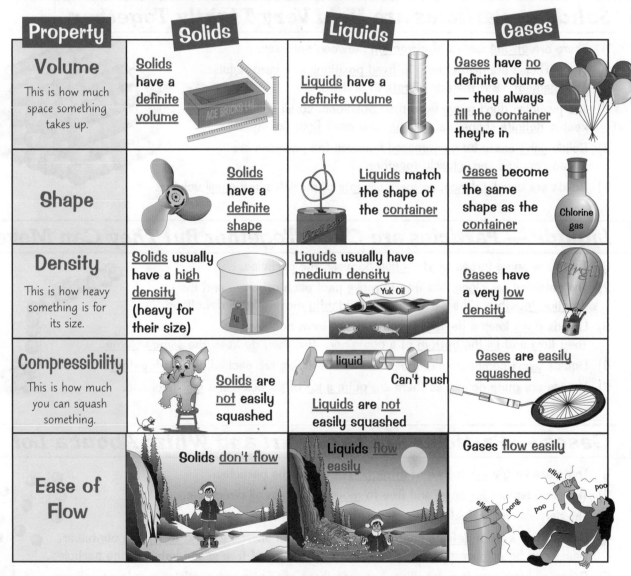

Property	Solids	Liquids	Gases
Volume — This is how much space something takes up.	<u>Solids</u> have a <u>definite volume</u>	<u>Liquids</u> have a <u>definite volume</u>	<u>Gases</u> have <u>no</u> definite volume — they always <u>fill the container</u> they're in
Shape	<u>Solids</u> have a <u>definite shape</u>	<u>Liquids</u> match the shape of the <u>container</u>	<u>Gases</u> become the same shape as the <u>container</u>
Density — This is how heavy something is for its size.	<u>Solids</u> usually have a <u>high</u> <u>density</u> (heavy for their size)	<u>Liquids</u> usually have <u>medium density</u>	<u>Gases</u> have a very <u>low</u> <u>density</u>
Compressibility — This is how much you can squash something.	<u>Solids</u> are <u>not</u> easily squashed	<u>Liquids</u> are <u>not</u> easily squashed	<u>Gases</u> are <u>easily</u> <u>squashed</u>
Ease of Flow	Solids <u>don't flow</u>	Liquids <u>flow easily</u>	Gases <u>flow easily</u>

The Three States — do they really matter....

<u>Solids</u>, <u>liquids</u> and <u>gases</u> — you must <u>learn</u> all those numbered points and the <u>properties</u> of all three. When you think you know it, <u>cover the page</u> and <u>scribble</u> it all down from <u>memory</u>. That soon shows what you really know — and what you <u>don't</u>. Keep trying till you can do it.

Particle Theory

Particle theory — sounds pretty <u>fancy</u>. But actually it's pretty <u>straightforward</u>.

1) The <u>particles</u> in a substance stay the <u>same</u> whether it's a <u>solid</u>, a <u>liquid</u> or a <u>gas</u>.

2) What changes is the <u>arrangement</u> of the particles and their <u>energy</u>.

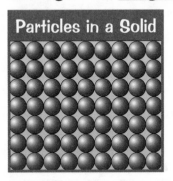
Particles in a Solid

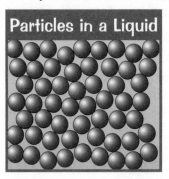
Particles in a Liquid

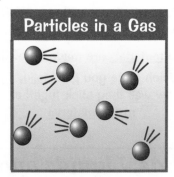
Particles in a Gas

3) This particle theory explains all the <u>different properties</u> of solids, liquids and gases...

Solids — *Particles are Held Very Tightly Together*

1) There are <u>strong</u> forces of <u>attraction</u> between particles.

2) The particles are held closely in <u>fixed positions</u> in a very regular <u>arrangement</u>. But they do <u>vibrate</u> to and fro.

3) The particles <u>don't move</u> from their positions, so all solids keep a <u>definite shape</u> and <u>volume</u>, and can't <u>flow</u> like liquids.

4) Solids <u>can't</u> easily be <u>compressed</u> because the particles are already packed <u>very closely together</u>.

5) Solids are usually <u>dense</u>, as there are <u>lots</u> of particles in a <u>small</u> volume.

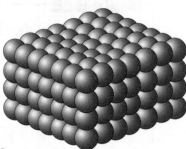

Liquids — *Particles are Close Together But They Can Move*

1) There are <u>some</u> forces of <u>attraction</u> between the particles.

2) The particles are <u>close</u>, but free to <u>move</u> past each other — and they do <u>stick together</u>. The particles are <u>constantly</u> moving in all directions.

3) Liquids <u>don't</u> keep a <u>definite shape</u> and can form puddles. They <u>flow</u> and <u>fill the bottom</u> of a container. But they do keep the <u>same volume</u>.

4) Liquids <u>won't</u> compress easily because the particles are packed <u>closely together</u>.

5) Liquids are <u>quite dense</u>, as there are <u>quite a lot</u> of particles in a <u>small</u> volume.

Gases — *Particles are Far Apart and Whizz About a Lot*

1) There are <u>very weak</u> forces of <u>attraction</u> between the particles.

2) The particles are <u>far apart</u> and free to <u>move</u> quickly in <u>all</u> directions.

3) The particles move <u>fast</u>, and so <u>collide</u> with each other and the <u>container</u>.

4) Gases <u>don't</u> keep a <u>definite shape</u> or <u>volume</u> and will always <u>expand to fill</u> any container. <u>Gases</u> can be <u>compressed easily</u> because there's a lot of free <u>space</u> between the particles.

5) Gases all have <u>very low densities</u>, because there are <u>not many</u> particles in a <u>large</u> volume.

Phew, Particle Theory — particle-ularly gripping stuff...

It's pretty <u>clever</u> the way you can explain all the differences between solids, liquids and gases with a page full of <u>snooker balls</u>. Anyway, that's the easy bit. The not-so-easy bit is making sure you <u>understand</u> it.

More Particle Theory

Particle theory can be used to explain all sorts of <u>exciting</u> things, like, erm, <u>gas pressure</u> and <u>diffusion</u>.
I say exciting. I may be <u>exaggerating</u> just a <u>teensy tiny</u> bit...

Gas Pressure *is Due to* Particles Hitting a Surface

Increasing *the* Temperature Increases Pressure

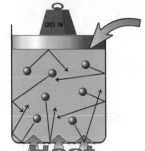

1) When you <u>increase</u> the <u>temperature</u>, it makes the particles move <u>faster</u>.
2) This has <u>two</u> effects:

> a) They hit the walls <u>harder</u>.
> b) They hit <u>more often</u>.

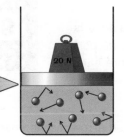

Increasing the temperature will only increase the pressure if the volume stays the same.

3) <u>Both</u> these things <u>increase</u> the <u>pressure</u>.

Reducing *the* Volume Increases Pressure

1) If you reduce the <u>volume</u> it makes the <u>pressure increase</u>.
2) This is because when the particles are <u>squashed up</u> into a <u>smaller space</u> they'll hit the walls <u>more often</u>.

Diffusion *is Just Particles* Spreading Out

1) Particles "want" to <u>spread out</u> — this is called <u>diffusion</u>.
An example is when a <u>smell</u> spreads slowly through a room.

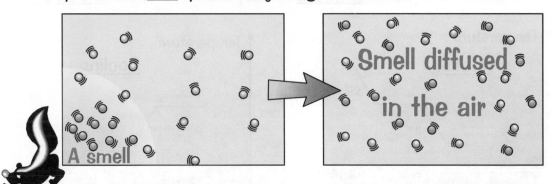

A smell

Smell diffused in the air

The smell particles <u>move</u> from an area of <u>high concentration</u> (i.e. where there are <u>lots of them</u>) to an area of <u>low concentration</u> (where there's <u>only a few</u> of them).

2) Diffusion is <u>slow</u> because the smell particles keep bumping into <u>air</u> particles, which stops them making forward progress and often sends them off in a completely different direction — it's a bit like trying to run blindfold through a herd of frisky bullocks. As you do.

Let the information on this page diffuse into your mind...

...you know — move from where there's a <u>high concentration</u> of information (this page) to where there's a <u>low concentration</u> (err... your mind). To be honest though, you're going to need to be a bit more <u>active</u> when it comes to <u>learning</u> this page. You know the drill by now. <u>Look</u>, <u>cover</u> and <u>scribble</u> it all down.

Physical Changes

Physical changes don't change the particles — just their arrangement or their energy.

Changes of *State* — i.e. changing from one state of matter to another.

3 At a certain temperature, the particles have enough energy to break free from their positions. This is called melting and the solid turns into a liquid.

4 When a liquid is heated, again the particles get even more energy.

2 This makes the particles move more which weakens the forces that hold the solid together. This makes the solid expand.

5 This energy makes the particles move faster which weakens the forces holding the liquid together.

6 At a certain temperature, the particles have enough energy to break the forces. This is called boiling and the liquid turns into a gas.

1 When a solid is heated, its particles gain more energy.

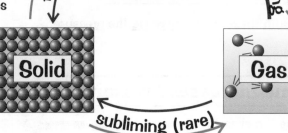

melting freezing

boiling condensing

subliming (rare)

A Red Arrow means energy is supplied

A Blue Arrow means energy is given out

A change of state doesn't involve a change in mass (see page 74), only a change in energy.

Heating **and** *Cooling Curves* **have** *Flat Bits*

Heating and cooling curves show the energy changes that happen when a substance changes state:

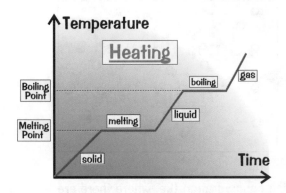

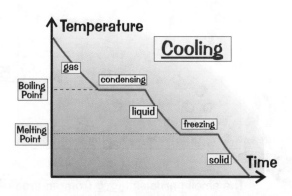

1) When a substance is melting or boiling, all the energy supplied from heating is used to weaken the forces between particles rather than raising the temperature — hence the flat bits on the heating graph.

2) When a substance is cooled, the cooling graph will show flat bits at the condensing and freezing points.

3) This is because the forces between particles get stronger when a gas condenses or when a liquid freezes — and energy is given out. This means that the temperature doesn't go down until all the substance has changed state.

Phew — another page of jostling snooker balls...

So the reason your ice cream melts is because the little snooker balls of ice cream take in energy, which means they can break free from their positions and become a liquid. Not that my ice creams last that long.

Atoms and Elements

If you've ever wondered what <u>everything is made of</u>, then the simple answer is <u>atoms</u>.

You Need to Know About Atoms...

1) Atoms are a type of <u>tiny</u>, <u>tiny</u>, <u>particle</u>.

2) They're so small that you <u>can't see them directly</u>.
   So for a long time, no one knew much about them.

3) <u>Dalton</u> was the first modern scientist to try to <u>explain</u> things about atoms.
   According to the <u>Dalton model</u>:

   - <u>All matter</u> is <u>made up</u> of <u>atoms</u>.
   - There are <u>different types of atom</u>.
   - Each <u>element</u> (see below) contains a <u>different type</u>.

   Scientists now know a lot more about atoms — but luckily, this is all you need to learn for Key Stage 3.

...and Elements

1) An <u>element</u> is a substance that contains <u>only one type</u> of <u>atom</u>.

2) Quite a lot of <u>everyday substances</u> are elements:

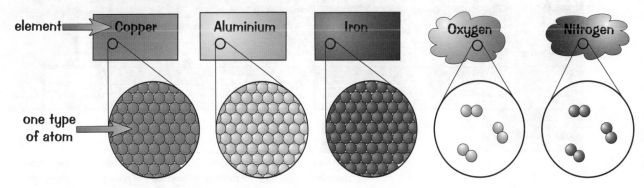

element → Copper   Aluminium   Iron   Oxygen   Nitrogen

one type of atom

3) All of these elements have <u>different properties</u>.
   For example, <u>copper</u> is a <u>soft</u>, <u>bendy metal</u>. <u>Oxygen</u> is a <u>colourless gas</u>.

All Elements Have a Name and a Symbol

1) There are over <u>100 different elements</u> and writing their names out each time you wanted to mention one would take ages.

2) So each element has a <u>symbol</u> — usually of <u>one or two letters</u>.

   Some symbols make sense (like O for oxygen) but others are based on Latin, so are a bit weird — like Fe for iron.

   He

   Examples:

   <u>Oxygen</u> has the symbol <u>O</u>.   <u>Helium</u> has the symbol <u>He</u>.
   <u>Carbon</u> has the symbol <u>C</u>.   <u>Iron</u> has the symbol <u>Fe</u>.

3) You can see the <u>symbol</u> for each element on the <u>periodic table</u> (see next page).

It's elemental, my dear Watson...

<u>Atoms</u> — you can't see them, but these tiny little footballs are absolutely <u>everywhere</u>.
Understanding <u>atoms</u> and <u>elements</u> is pretty <u>fundamental</u> to the whole of chemistry — so make
absolutely certain <u>you've learnt this page</u> before you even think about moving on to the next one.

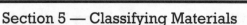

The Periodic Table

How can you find out what all those <u>chemical symbols</u> mean? Step forth the periodic table...

The Periodic Table Lists All the Elements

1) The periodic table shows all the <u>elements</u> we have <u>discovered</u>.

2) The <u>first version</u> of the table was put together by a scientist called <u>Mendeleev</u>. It's thanks to Mendeleev that <u>elements</u> with <u>very similar properties</u> are arranged into <u>vertical columns</u> in the table.

3) The <u>vertical columns</u> are called <u>groups</u>.

4) The <u>horizontal rows</u> are called <u>periods</u>.

5) If you know the <u>properties</u> of <u>one element</u> in a <u>group</u>, you can <u>predict</u> the properties of <u>other elements</u> in that group. E.g. <u>Group 1</u> elements are all <u>soft</u>, <u>shiny metals</u>, which react in a similar way with <u>water</u>.

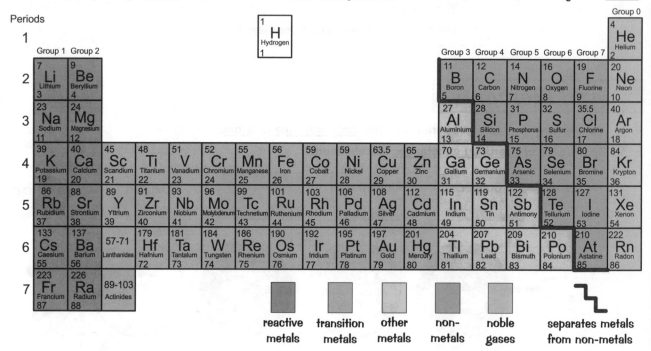

reactive metals   transition metals   other metals   non-metals   noble gases   separates metals from non-metals

You Can Use the Periodic Table to Predict Patterns in Reactions

1) In a chemical reaction, <u>elements combine</u> to form new substances (see next page).

2) An element that's <u>dead keen</u> to combine with other elements is said to be very <u>reactive</u>. <u>Group 1</u>, <u>2</u> and <u>7</u> elements are all <u>pretty reactive</u>.

3) <u>Group 0</u> elements (the "<u>noble gases</u>") are all <u>extremely unreactive</u>. They <u>almost never</u> take part in <u>any</u> chemical reactions.

4) You can use the periodic table to <u>predict patterns</u> in chemical reactions. For example...

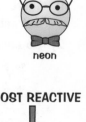

neon

The <u>Group 1</u> metals get <u>MORE</u> <u>reactive</u> as you go <u>down</u> the group. You can <u>see</u> this by the way the Group 1 metals <u>react with water</u>. When <u>lithium</u> (<u>Li</u>) reacts with water, it <u>fizzes</u>. When <u>rubidium</u> (<u>Rb</u>) reacts with water, it <u>explodes</u>. This is because rubidium is <u>much more reactive</u> than lithium.

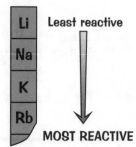

The non-metals in <u>Group 7</u> behave in the <u>opposite</u> way to the metals in Group 1. They get <u>LESS</u> <u>reactive</u> as you go <u>down</u> the group.

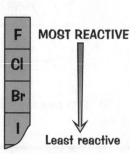

Group 1 metals — always spoiling for a fight...

So... the periodic table is made up of <u>groups</u> and <u>periods</u>. All the elements in a <u>group</u> have <u>similar properties</u>.

Compounds

It'd be pretty boring if we only had elements to play with,
but luckily <u>compounds</u> give us all sorts of <u>exciting materials</u>...

Compounds *Contain* Two or More *Elements* Joined Up

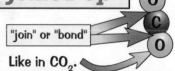

1) When <u>two or more atoms join</u> together, a <u>molecule</u> is made. The "<u>join</u>" is known as a <u>chemical bond</u>.

2) <u>Compounds</u> are formed when atoms from <u>different elements</u> join together. Like in CO_2.

"join" or "bond"

An <u>ELEMENT</u> which is made up of <u>atoms</u>	An <u>ELEMENT</u> which is made up of <u>molecules</u>	Molecules in a <u>COMPOUND</u>	A <u>MIXTURE</u> of different <u>elements</u>
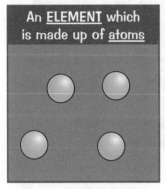	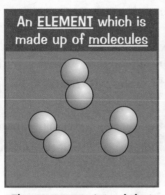	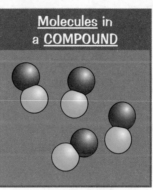	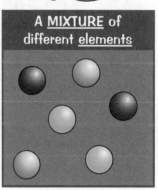
The atoms are all the same and not joined up — it must be an <u>element</u>.	The atoms are joined, but there's still only one type, so it's still an <u>element</u>.	Here we have different atoms joined together — that's a <u>compound</u> alright.	This is <u>not</u> a compound because the elements aren't joined up — it's a <u>mixture</u>.

Compounds *are Formed from* Chemical Reactions

1) A <u>chemical reaction</u> involves chemicals (called the <u>reactants</u>) <u>combining</u> together or <u>splitting</u> apart to form one or more <u>new</u> substances (called the <u>products</u>).

2) When a <u>new</u> compound is <u>synthesised</u> (made), elements <u>combine</u>.

3) The <u>new compounds</u> produced by any chemical reaction are always totally <u>different</u> from the <u>original elements</u> (or reactants). The <u>classic example</u> of this is <u>iron</u> reacting with <u>sulfur</u> as shown below:

Iron is <u>magnetic</u>. It reacts with <u>sulfur</u> to make <u>iron sulfide</u>, a totally new substance which is <u>not magnetic</u>.

These <u>equations</u> show what happens in the reaction:

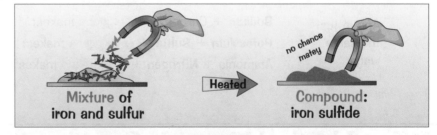

Mixture of iron and sulfur

Heated

no chance matey

Compound: iron sulfide

Word equation: Iron + Sulfur [Heated] Iron Sulfide

In symbols: Fe + S [Heated] FeS

4) When elements undergo a <u>chemical reaction</u> like the one above, the products will always have a <u>chemical formula</u> — e.g. H_2O for <u>water</u> or <u>FeS</u> for <u>iron sulfide</u>.

5) Compounds can be <u>split up</u> back into their <u>original</u> elements but it <u>won't</u> just happen by itself — you have to <u>supply</u> a lot of <u>energy</u> to make the reaction go in <u>reverse</u>.

Remember, every element has a name and a symbol. See p. 35 for more.

Learn about Compounds — and try and make it stick...

They really do like seeing if you know the difference between <u>elements</u> and <u>compounds</u> (and mixtures — see page 39 for more on these). Make sure you do. You'll also need to learn what happens in <u>chemical reactions</u>. It's not that tricky — but you do have to <u>learn</u> all the <u>picky details</u> on this page good 'n' proper.

38

Naming Compounds

When elements combine to make a compound, their names change slightly. Learn the Two Simple Rules.

Naming Compounds — *Two Simple Rules*

Rule 1: When two different elements combine the ending is usually "something -ide".

 NaCl ⟵ Formula ⟶ MgO

Sodium and Chlorine give: SODIUM CHLORIDE

⟵ Elements present ⟶
⟵ Name of Compound ⟶

Magnesium and Oxygen give: MAGNESIUM OXIDE

And in just the same way:

Sulfur changes to Sulfide
Iodine changes to Iodide

Bromine changes to Bromide
Fluorine changes to Fluoride

Rule 2: When three or more different elements combine — and one of them is oxygen — the ending will usually be "something -ate".

 CuSO$_4$ ⟵ Formula ⟶ CaCO$_3$

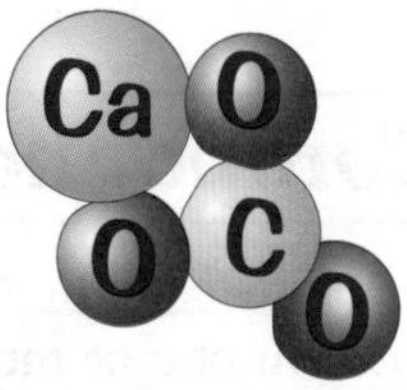

1 Copper, 1 Sulfur, 4 Oxygens
COPPER SULFATE

⟵ Elements present ⟶
⟵ Name of Compound ⟶

1 Calcium, 1 Carbon, 3 Oxygens
CALCIUM CARBONATE

And in just the same way:

Sodium + Carbon + 3 Oxygens makes: SODIUM CARBONATE
Potassium + Sulfur + 4 Oxygens makes: POTASSIUM SULFATE
Ammonia + Nitrogen + 3 Oxygens makes: AMMONIUM NITRATE

If *Two Identical* Elements Combine, *it's Not a* Compound

Identical atoms of the same element are often found combined.
This doesn't make them a compound though — in fact, their name doesn't even change.

H$_2$ = Hydrogen (H H)
N$_2$ = Nitrogen (N N)
O$_2$ = Oxygen (O O)

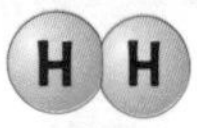

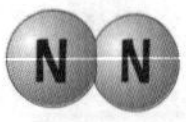

F$_2$ = Fluorine (F F)
Cl$_2$ = Chlorine (Cl Cl)
Br$_2$ = Bromine (Br Br)

These are all elements with two atoms, not compounds. They're almost never found as single atoms in nature.

-ides and -ates — it's the game of the name, pal...

Naming compounds can get very tricky of course, but for KS3 you'll do just fine learning those two simple rules. When you have, work out the names of these three:
1) Sodium joined with fluorine   2) Two iodines joined up   3) Calcium with sulfur and oxygen.*

Section 5 — Classifying Materials

*Answers on page 108

Mixtures

Mixtures in chemistry are like <u>cake mix</u> in the kitchen — all the components are <u>mushed up</u> together, but you can still <u>pick out</u> the raisins if you really want. You'll need to learn the technical terms too though...

Mixtures are Substances That are NOT Chemically Joined Up

1) A <u>pure substance</u> is made up of only <u>one type</u> of <u>element</u> OR only <u>one type</u> of <u>compound</u>. It <u>can't</u> be <u>separated</u> into anything simpler without a <u>chemical reaction</u>.

> E.g. <u>pure water</u> is made up of <u>H_2O molecules only</u>. These molecules can't be separated into H and O atoms <u>without</u> a chemical reaction.

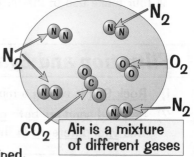

2) A <u>mixture</u> contains <u>two</u> or more <u>different substances</u>. These substances aren't chemically joined up — so, if you're clever, you can <u>separate</u> them very <u>easily</u> using <u>physical methods</u> (i.e. without a chemical reaction). See pages 40-41 for more.

3) <u>Sea water</u> and <u>air</u> are good <u>examples</u> of mixtures — they contain several different substances which aren't chemically combined.

Air is a mixture of different gases

4) A mixture has the <u>properties</u> of <u>its constituent parts</u> (i.e. the parts it's made from).

Dissolving isn't Disappearing

1) <u>Dissolving</u> is a common way mixtures are made.

2) When you add a solid (the <u>solute</u>) to a liquid (the <u>solvent</u>) the <u>bonds</u> holding the solute particles together sometimes <u>break</u>.

3) The solute particles then <u>mix</u> with the particles in the liquid forming a <u>solution</u>.

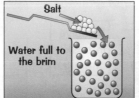 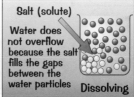 

<u>Learn</u> these seven definitions:

1) <u>Solute</u> – is the solid being dissolved.
2) <u>Solvent</u> – is the liquid it's dissolving into.
3) <u>Solution</u> – is a mixture of a solute and a solvent that does not separate out.
4) <u>Soluble</u> – means it <u>WILL</u> dissolve.

5) <u>Insoluble</u> – means it will <u>NOT</u> dissolve.
6) <u>Saturated</u> – a solution that won't dissolve any more solute at that temperature.
7) <u>Solubility</u> – a measure of how much solute will dissolve.

4) Remember, when salt <u>dissolves</u> it hasn't <u>vanished</u> — it's still <u>there</u> — <u>no mass</u> is lost.

5) If you <u>evaporated</u> off the <u>solvent</u> (the water), you'd see the <u>solute</u> (the salt) again.

20g Salt   Added to 100g Water = 120g Solution

Solubility Increases with Temperature

1) At <u>higher</u> temperatures <u>more solute</u> will dissolve in the <u>solvent</u> because particles move faster.

2) However <u>some</u> solutes won't dissolve in certain <u>solvents</u>. E.g. salt won't dissolve in petrol.

I said nothing disappears — just don't ask Derren Brown...

<u>Learn</u> the three main headings on this page till you can write them down <u>from memory</u>. Then <u>learn</u> the stuff underneath them, including the diagrams. <u>Cover the page</u> and <u>write it all down</u>.

Separating Mixtures

There are all sorts of ways you can separate mixtures. You've got to know <u>four</u> of them.

*Mixtures **Can be** Separated **Using** Physical Methods*

There are <u>four separation techniques</u> you need to be familiar with.

1) <u>FILTRATION</u>   2) <u>EVAPORATION</u>   3) <u>CHROMATOGRAPHY</u>   4) <u>DISTILLATION</u> (see next page).

All four make use of the <u>different properties</u> of the <u>constituent parts</u> to <u>separate</u> them out.

*Filtration **and** Evaporation — E.g. for the Separation of Rock Salt*

1) <u>Rock Salt</u> is simply a <u>mixture</u> of <u>salt</u> and <u>sand</u> (they spread it on the roads in winter).

2) Salt and sand are both <u>compounds</u> — but <u>salt dissolves</u> in water and <u>sand doesn't</u>. This <u>vital difference</u> in their <u>physical properties</u> gives us a great way to <u>separate</u> them.

You Need to Learn the Four Steps of the Method:

1) <u>Grinding</u>   2) <u>Dissolving</u>   3) <u>Filtering</u>   4) <u>Evaporating</u>

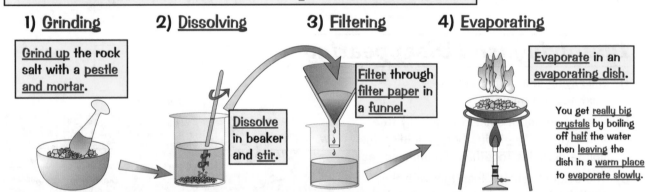

Grind up the rock salt with a <u>pestle and mortar</u>.

Dissolve in beaker and <u>stir</u>.

Filter through filter paper in a funnel.

Evaporate in an evaporating dish.

You get <u>really big crystals</u> by boiling off <u>half</u> the water then <u>leaving</u> the dish in a <u>warm place</u> to <u>evaporate slowly</u>.

• The sand doesn't dissolve (it's <u>insoluble</u>) so it stays as <u>big grains</u> and obviously these <u>won't fit</u> through the <u>tiny holes</u> in the filter paper — so it <u>collects on the filter paper</u>.

• The <u>salt</u> is dissolved in <u>solution</u> so it does go through — and when the water's <u>evaporated</u>, the salt forms as <u>crystals</u> in the <u>evaporating dish</u>. This is called <u>crystallisation</u>. (Surprise surprise.)

*Chromatography **is Ideal for Separating** Dyes **in Inks***

1) <u>Different dyes</u> in ink will <u>wash</u> through paper at <u>different rates</u>.

2) Some will <u>stick</u> to the <u>paper</u> and others will <u>dissolve</u> in the <u>solvent</u> (see below) and <u>travel</u> through it <u>quickly</u>.

Chromatography can also be used to identify blood samples and investigate chlorophyll.

Method 1

1) <u>Dots of ink</u> are put onto <u>filter paper</u>.

2) A <u>wick</u> is cut from part of the paper (as shown).

3) The <u>solvent</u> washes the <u>dyes</u> through the paper.

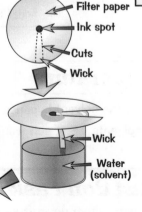

Filter paper, Ink spot, Cuts, Wick, Water (solvent)

Different dyes in the ink

Method 2

1) Put <u>spots</u> of <u>inks</u> onto a pencil <u>baseline</u> on <u>chromatography paper</u>.

2) <u>Roll</u> the sheet up and put it in a <u>beaker</u>.

3) The solvent <u>seeps</u> up the paper, carrying the ink dyes with it.

4) Each different dye will form a <u>spot</u> in a different place.

5) You can <u>compare</u> a forged ink to a <u>known ink</u> to see which it is.

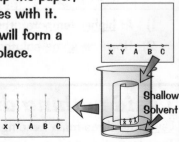

X Y A B C   Shallow Solvent

Separating Mixtures

Simple *Distillation* Separates *Pure Water* from *Ink*

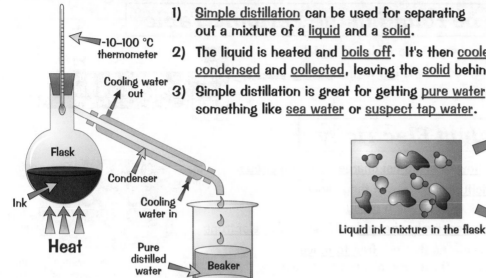

1) <u>Simple distillation</u> can be used for separating out a mixture of a <u>liquid</u> and a <u>solid</u>.

2) The liquid is heated and <u>boils off</u>. It's then <u>cooled</u>, <u>condensed</u> and <u>collected</u>, leaving the <u>solid</u> behind.

3) Simple distillation is great for getting <u>pure water</u> from something like <u>sea water</u> or <u>suspect tap water</u>.

-10–100 °C thermometer

Cooling water out

Flask

Condenser

Ink

Cooling water in

Heat

Pure distilled water

Beaker

Concentrated ink in the flask

Liquid ink mixture in the flask

Pure water in the beaker

Fractional *Distillation* Separates *Mixed* Liquids

1) <u>Fractional distillation</u> is used for separating a mixture of liquids like <u>crude oil</u>.

2) <u>Different liquids</u> will boil off at <u>different temperatures</u>, around their <u>own boiling point</u>.

3) The <u>fractionating column</u> ensures that the "wrong" liquids <u>condense</u> back <u>down</u>, and only the liquid properly <u>boiling</u> at the temperature on the thermometer will make it to the top.

4) When each liquid has <u>boiled off</u>, the temperature reading <u>rises</u> until the <u>next</u> fraction starts to boil off.

5) Real life <u>examples</u> include:
   * distilling <u>whisky</u>,
   * separating <u>crude oil</u> into petrol, diesel and other fuels.

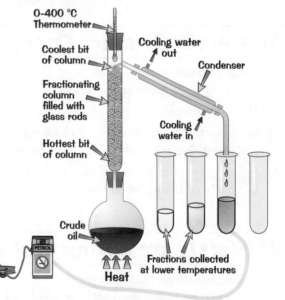

0-400 °C Thermometer

Coolest bit of column

Cooling water out

Condenser

Fractionating column filled with glass rods

Cooling water in

Hottest bit of column

Crude oil

PETROL

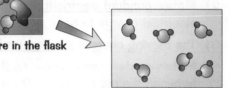

Fractions collected at lower temperatures

Heat

Check *Purity* with *Melting* and *Boiling Points*

1) A <u>pure</u> chemical substance has <u>fixed melting and boiling points</u>. E.g. pure water boils at 100 °C and pure ice melts at 0 °C. These figures are <u>known</u> for a huge range of substances.

2) This helps us to <u>identify unknown</u> substances, e.g. if a liquid boils at <u>exactly</u> 100 °C it's likely to be <u>pure</u> water.

3) <u>Impurities change</u> melting and boiling points, e.g. impurities in water cause it to boil <u>above</u> 100 °C.

4) This means you can <u>test the purity</u> of a substance you've separated from a mixture.

Pure Substance	Melting Point °C	Boiling Point °C
Water	0	100
Ethanol	-114	78
Aluminium	660	2520

Revise mixtures — just filter out the important bits...

Teachers love asking you about separation techniques (strange, I know), so make sure you've got all the <u>facts</u> AND the <u>diagrams</u> absolutely 100% learnt. Make sure you know how to check <u>purity</u> too.

Properties of Metals

Metals are jolly <u>useful</u>. We use them all the time in <u>bendy wires</u>, <u>bridges</u>, <u>musical instruments</u> and more. So it's only fair that you learn these two pages of glorious <u>facts</u> about them in return...

1) Metals *Can be Found in the* Periodic Table

1) <u>Most</u> of the elements in the periodic table are metals.
2) Some are shown here in red, to the <u>left</u> of the <u>zig zag</u>. ➡

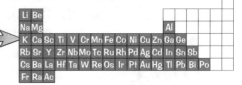

2) Metals **Conduct** *Electricity*

1) <u>Electric current</u> is the <u>flow</u> of <u>electrical charge</u> around a <u>circuit</u>.
2) Metals <u>conduct electricity</u>, which means they allow electrical charge to pass through them <u>easily</u>.
3) The moving charges are <u>negatively-charged</u> particles called <u>electrons</u>.
4) Metals contain some electrons that are <u>free to move</u> between the metal atoms. These free electrons can carry an electric current from <u>one end</u> of the metal to the <u>other</u>.
5) Because they conduct electricity well, metals are often used to make <u>wires</u> and parts of <u>electrical circuits</u>.

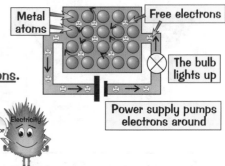

3) Metals **Conduct** *Energy*

1) Metals transfer energy from a <u>hot place</u> to a <u>cold place</u> quickly and easily.
2) The <u>"hot"</u> particles <u>vibrate strongly</u>.
3) Because the particles are very close together, the vibrations are easily <u>passed on</u> through the metal.
4) <u>Free electrons</u> in the metal also help to <u>transfer energy</u> from the <u>hot</u> parts of the metal to the <u>cooler</u> parts as they move around.

CONDUCTION OF ENERGY

<u>Lots of movement</u>   <u>Little movement</u>

4) Metals *are Strong and Tough*

1) Metals have high <u>tensile strength</u> (they can be pulled hard without breaking).
2) This is because there are <u>strong forces</u> between metal atoms that <u>hold them together</u>. Their strength makes them good <u>building materials</u>.

5) Metals *are Shiny **When** Polished*

<u>Polished</u> or <u>freshly cut</u> metals give strong <u>reflection</u> of light from their <u>smooth surface</u>. This makes them look <u>shiny</u>.

6) Metals *are Sonorous*

This means they make a nice <u>"donnnngggg"</u> sound when they're hit. If you think about it, it's <u>only metals</u> that do that — you <u>could</u> make a gong out of plastic, but it wouldn't be much good.

Section 5 — Classifying Materials

Properties of Metals

7) Metals are Malleable

1) Metals are <u>easily shaped</u> (malleable) because the atoms in metals can <u>slide over</u> each other.

2) This means metals can be <u>hammered</u> into <u>thin sheets</u> or <u>bent</u> — all <u>without shattering</u>.

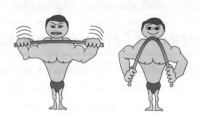

8) Metals are Ductile

1) This means they can be drawn into <u>wires</u>.

2) Metals <u>aren't brittle</u> like non-metals (see page 45) are. They just <u>bend</u> and <u>stretch</u>.

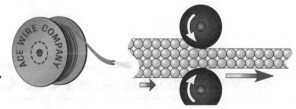

9) Metals have High Melting and Boiling Points

1) A <u>lot</u> of <u>energy</u> is needed to <u>melt</u> metals.

2) This is because their <u>atoms</u> are joined by <u>strong</u> forces.

3) The table shows how hot they have to get to <u>melt</u>.

Metal	Melting Point (°C)	Boiling Point (°C)
Aluminium	660	2520
Copper	1085	2562
Magnesium	650	1090
Iron	1538	2861
Zinc	420	907
Silver	962	2162

10) Metals have High Densities

1) <u>Density</u> is all to do with how much <u>stuff</u> there is squeezed into a certain <u>space</u>.

2) Metals feel <u>heavy</u> for their <u>size</u> (i.e. they're <u>very dense</u>) because they have a lot of <u>atoms</u> tightly packed into a <u>small volume</u>.

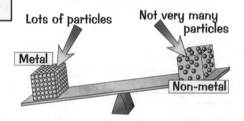

Lots of particles

Not very many particles

Metal

Non-metal

11) Metals Make Alloys When Mixed with Other Metals

1) A <u>combination</u> of different metals is called an <u>alloy</u>. The <u>properties</u> of the metals get <u>jumbled up</u> in the new <u>alloy</u>.

2) So <u>lighter, weaker metals</u> can be <u>mixed</u> with <u>heavier, stronger metals</u> and the <u>result</u> is, hopefully, an <u>alloy</u> which is <u>light and strong</u>.

Alloy Wheels — light and strong

12) Some Metals are Magnetic

1) Only <u>certain metals</u> are magnetic.

2) <u>Most</u> metals <u>aren't magnetic</u>. <u>Iron</u>, <u>nickel</u> and <u>cobalt</u> are. <u>Alloys</u> made with these three metals will also be magnetic — e.g. <u>steel</u> is made mostly from <u>iron</u>, so is also <u>magnetic</u>.

Iron or nickel or cobalt (or an alloy containing one of them)

Good Alloys — you can rely on friends from Birmingham...

There they are then. A whole load of facts about metals just waiting to be soaked up into that giant sponge lurking between your ears. You need to <u>keep practising</u> till you can <u>scribble down all the headings</u> with both pages <u>covered</u>. Then try <u>filling in the details</u>. Then turn your doodles into an awesome paper aeroplane.

Properties of Non-Metals

The properties of non-metal elements vary quite a lot. Good — life would stink if everything was like sulfur...

1) Non-metals Can be Found in the Periodic Table

1) All the non-metals (with the exception of hydrogen) are clustered in the corner over on the right of the zig zag. Look, right over there.

2) There are fewer non-metals than metals.

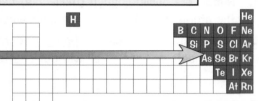

2) Non-metals are Poor Conductors of Electricity

1) Most non-metals are insulators, which means that charges can't flow through them.

2) If charges can't move then no electric current flows. This is very useful — non-metals combine to make things like plugs and electric cable coverings.

> One exception to this rule is graphite — a non-metal made purely from carbon atoms. Its atoms are arranged in layers, which allow electrons to move along them, so graphite can conduct electricity.

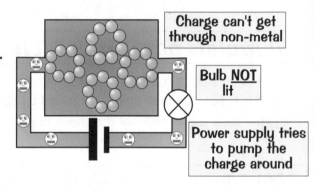

Charge can't get through non-metal

Bulb **NOT** lit

Power supply tries to pump the charge around

3) Non-metals are Poor Conductors of Energy by Heating

1) Non-metals don't transfer energy from a hot place to a cold place very quickly or easily.

2) This makes non-metals really good insulators.

3) "Hot" particles don't pass on their vibrations so well.

You can't travel on this bus pal!

Well I'm going to find a decent conductor

The Non-Metal Bus

4) Non-metals are NOT Strong or Hard-Wearing

1) The forces between the particles in non-metals are weak — this means they break easily.

2) It's also easy to scrub atoms or molecules off them — so they wear away quickly.

This carbon brush keeps wearing out

Properties of Non-Metals

5) *Non-metals* **are** *Dull*

1) Most non-metals don't <u>reflect</u> light very well at all. Their surfaces are not usually as <u>smooth</u> as metals.

2) This makes them look <u>dull</u>.

6) *Non-metals* **are** *Brittle*

1) Non-metal <u>structures</u> are held together by <u>weak forces</u>.

2) This means they can <u>shatter</u> all too easily.

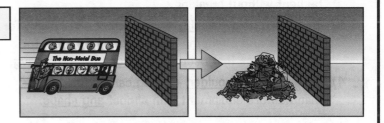

7) *Non-metals* **Have** *Low Melting Points* **and** *Boiling Points*

1) The <u>forces</u> which hold the particles in non-metals <u>together</u> are <u>very weak</u>. This means they <u>melt</u> and <u>boil</u> very <u>easily</u>.

2) At room temperature, most non-metals are <u>gases</u> or <u>solids</u>. Only one is a liquid.

Non-Metal	Melting Point (°C)	Boiling Point (°C)
Sulfur	113	445
Oxygen	-218	-183
Chlorine	-101	-35
Helium	-272	-269
Neon	-249	-246
Bromine	-7	59

8) *Non-metals* **Have** *Low Densities*

1) Obviously the non-metals which are <u>gases</u> will have <u>very low density</u>. Some of these gases will even <u>float</u> in <u>air</u> — ideal for party balloons.

2) This means they don't have very many <u>particles</u> packed into a certain <u>space</u>.

3) Even the liquid and solid non-metals have <u>low densities</u>.

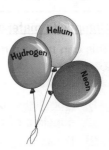

9) *Non-metals* **are** *Not Magnetic*

1) Only a few <u>metals</u> like <u>iron</u>, <u>nickel</u> and <u>cobalt</u> are <u>magnetic</u>.

2) <u>All non-metals</u> are most definitely <u>non-magnetic</u>.

Non-Metals — they REALLY ARE dull aren't they...

You still have to learn all about them though. Do it like this: <u>Cover the page</u> with a bit of paper and try and <u>write down</u> each of the 9 points, one at a time. Lower the paper each time to see if you scribbled it all down right. Keep trying <u>till you can get them all</u>. Then put on some nice relaxing music by Non-Metallica.

Properties of Other Materials

As well as metals and non-metals, you need to learn all about some ace compounds and mixtures of compounds — polymers (plastics), ceramics (like bone china) and composites (like fibreglass).

Polymers *Have Many* Useful Properties

Polymers (that's plastics to you and me) include nylon, polythene and PVC.

1) Polymers are usually insulators — it's difficult for energy to be transferred through them electrically or by heating.

2) They're often flexible — they can be bent without breaking.

3) They have a low density — they can be very light for their size and strength. This makes them ideal for making things that need to be strong but not heavy.

4) They're easily moulded — they can be used to manufacture equipment with almost any shape.

Polymers are used to make everything from crash helmets and kayaks to carrier bags and drinks bottles.

Polymers are just compounds, made by joining loads of little molecules together in long chains. They usually contain carbon.

Ceramics *are Stiff but Brittle*

Ceramics include glass, porcelain and bone china (for posh tea cups). They are:

1) Insulators — it's difficult for energy to be transferred through them electrically or by heating.

2) Brittle — they aren't very flexible and will break instead of bending.

3) Stiff — they can withstand strong forces before they break.

As well as tea cups, ceramics are used for brakes and parts of spark plugs in cars.

Ceramics are made by 'baking' substances like clay.

Composites *are Made of Different Materials*

1) Composite materials are made from two or more materials stuck together.

2) This can make a material with more useful properties than either material alone. For example:

Fibreglass

1) Fibreglass (or Glass Reinforced Plastic — GRP) consists of glass fibres embedded in plastic.

2) It has a low density (like plastic) but is very strong (like glass).

3) These properties mean fibreglass is used for things like skis, boats and surfboards.

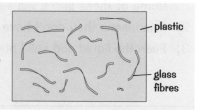

plastic

glass fibres

Concrete

1) Concrete is made from a mixture of sand and gravel embedded in cement.

2) It can withstand high compression stresses (i.e. being squashed) so it's great at supporting heavy things. This makes it ideal for use as a building material, e.g. in skate parks, shopping centres, airports, etc.

Boredom is a common property of revision...

Polymers, ceramics and composites. They're pretty handy — especially when the alternative is a wooden surfboard. Now you need to master everything on this page until you can describe all three AND their uses without any sneaky peeking. Once that's done, stare smugly out the window for 30 seconds.

Section Summary

We've moved on to Chemistry now. Makes a refreshing change from all that slimy Biology anyway.
Section 5 is all about Classifying Materials so here's a whole page of delicious Section Summary questions to help you classify how much you've remembered.

You know the drill: work through these questions and try to answer them. For any you can't do, look back through Section 5 and find the answer — and then learn it. Then try all the questions again and see how many more you can do that time. Keep going and before you know it you'll be answering them all perfectly.

1) What are the three states of matter? Describe five properties for each of them.
2) Draw what the particles look like in a solid, a liquid and a gas.
3) Explain how gases exert a pressure on the insides of a container.
4) What happens to the pressure of a gas if the temperature of the gas is increased?
5) What happens to the pressure of a gas if the volume of the gas is decreased?
6) Explain what diffusion is.
7) Give the names of five changes of state, and say which state they go from and to.
8) For any given substance, in which state do the particles have the most energy?
9) Does a change of state involve a change in mass?
10) Explain why a heating curve has a flat bit when a substance is boiling.
11) What is an atom?
12) What is an element? Roughly how many elements are there in the periodic table?
13) In the periodic table: a) What is a group?   b) What is a period?
14) Using the periodic table, give the chemical symbol for these:
    a) sodium   b) magnesium   c) oxygen   d) iron   e) sulfur
    f) aluminium   g) carbon   h) chlorine   i) calcium   j) zinc.
15) Which will show a more violent reaction with water — lithium or rubidium? Why?
16) Use the periodic table to predict whether fluorine or iodine is more reactive.
17) What is a compound? How is a compound different from a mixture?
18) Sketch some molecules that could be in a compound.
19) In what way is iron sulfide different from a mixture of iron and sulfur?
20) Is it easy to split a compound back up into its original elements?
21) Write down the two rules for naming compounds.
22) If two atoms of the same elements combine, what happens to their name?
23)*Give the name of the following:
    a) MgO   b) CaO   c) NaCl   d) $CaCO_3$ e) $CuSO_4$
24)*Give the name of the compound you get from chemically joining up these:
    a) sodium with chlorine   b) magnesium with chlorine   c) magnesium with carbon and oxygen.
25) What is a pure substance? What is a mixture?
26) Describe what happens when a substance dissolves.
27) What happens to solubility when the temperature increases?
28) List four mixture separation techniques with an example for each one.
29) Which of them would you use to try to identify different colours in a paint?
30) List the 12 facts you need to know about metals.
    Then list the 9 facts you need to know about non-metals.
31) Out of metals and non-metals, which are the:
    a) best conductors   b) most brittle   c) strongest   d) best insulators?
32) Name 4 useful properties of polymers.
33) What are ceramics useful for?
34) What are composites? Name one and describe what it's made of.

*Answers on page 108

Section 6 — Chemical Changes

Chemical Reactions

In a chemical reaction, all that's really happening is the <u>atoms moving around</u> into new formations. The reactants might give off energy, make a loud bang, or do a little dance, but the <u>mass won't change</u>.

Atoms Rearrange *Themselves in a* Chemical Reaction

1) In a <u>chemical reaction</u> atoms are <u>not</u> created or destroyed.

2) The atoms at the <u>start</u> of a reaction are <u>still there</u> at the <u>end</u>.

3) <u>Bonds</u> get <u>broken</u> and <u>made</u> in the reaction, as atoms <u>rearrange</u> themselves in going from the <u>reactants</u> to the <u>products</u> (p. 37). But the atoms themselves are <u>not altered</u>.

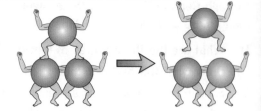

EXAMPLE: zinc + copper sulfate ⟶ zinc sulfate + copper

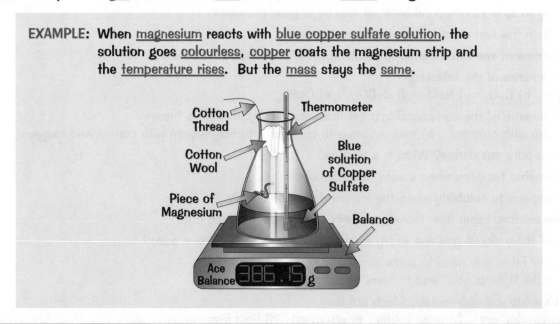

The *Mass Doesn't Change in a* Chemical Reaction

1) In a chemical reaction <u>no mass</u> is <u>lost</u> or <u>gained</u> when the <u>reactants</u> turn into the <u>products</u>.

2) This is because the <u>total number</u> of <u>atoms</u> is the <u>same</u> before and after the reaction.

3) Chemical reactions involve a change in <u>energy</u>, i.e. reactions always <u>give out</u> or <u>take in</u> energy (p. 50). This energy is usually <u>transferred</u> by <u>heating</u>, which causes the <u>temperature</u> in a reaction to go up or down.

4) <u>Visible changes</u> can occur in the reaction mixture. These show that a reaction has taken place. For example — a <u>gas</u> comes off, a <u>solid</u> is made, or the <u>colour</u> changes.

EXAMPLE: When <u>magnesium</u> reacts with <u>blue copper sulfate solution</u>, the solution goes <u>colourless</u>, <u>copper</u> coats the magnesium strip and the <u>temperature rises</u>. But the <u>mass</u> stays the <u>same</u>.

Cotton Thread
Thermometer
Cotton Wool
Blue solution of Copper Sulfate
Piece of Magnesium
Balance
Ace Balance 386.15 g

Chemical reactions — just a case of atomic acrobatics...

Some <u>chemical reactions</u> involve colour changes, heating up, stinky emissions and even explosions, but there's one thing that always stays the <u>same</u> — the total <u>mass</u>, before and after the reaction. Get your head round that and you've practically solved all of KS3 chemistry. Okay, maybe that's a slight exaggeration.

Examples of Chemical Reactions

Three common examples of <u>chemical reactions</u> coming right up... just what the doctor ordered.

Combustion *is Burning* in Oxygen

1) Combustion is <u>burning</u> — a <u>fuel</u> reacts with <u>oxygen</u> to release <u>energy</u>.

2) <u>Three</u> things are needed for combustion:
   1) Fuel
   2) Heating
   3) Oxygen

3) <u>Hydrocarbons</u> are <u>fuels</u> containing only <u>hydrogen</u> and <u>carbon</u>. When it's <u>hot</u> enough there's enough <u>oxygen</u>, hydrocarbons <u>combust</u> (burn) to give <u>water</u> and <u>carbon dioxide</u>:

$$\text{hydrocarbon} + \text{oxygen} \longrightarrow \text{carbon dioxide} + \text{water} (+ \text{energy})$$

4) Combustion is useful because <u>energy</u> is transferred away by <u>heating</u> and <u>light</u>. It's the process behind candles, wood fires, car engines, coal power plants, etc.

Oxidation *is the* Gain *of* Oxygen

1) When a substance <u>reacts</u> and <u>combines</u> with <u>oxygen</u>, it's called an <u>oxidation</u> reaction.

2) <u>Combustion</u> is an oxidation reaction.

3) Another example of oxidation is <u>rusting</u>. <u>Iron</u> reacts with <u>oxygen</u> in the air to form <u>iron oxide</u>, i.e. <u>rust</u>.

$$\text{iron} + \text{oxygen} \longrightarrow \text{iron oxide (rust)}$$

Thermal Decomposition *is Breaking Down* When Heated

1) <u>Thermal decomposition</u> is when a substance <u>breaks down</u> into at least two other substances when <u>heated</u>.

2) The substance <u>isn't</u> actually <u>reacting</u> with anything, but it <u>is</u> a <u>chemical</u> change.

3) Some <u>metal carbonates</u> break down on heating. Carbonates are substances with CO_3 in them, like copper(II) carbonate ($CuCO_3$) and zinc carbonate ($ZnCO_3$).

4) They break down into a <u>metal oxide</u> (e.g. copper oxide, CuO) and <u>carbon dioxide</u>. This usually results in a <u>colour change</u>.

EXAMPLE: The thermal decomposition of copper(II) carbonate.

copper(II) carbonate $\longrightarrow$ copper(II) oxide + carbon dioxide
$CuCO_3$ CuO   + CO_2

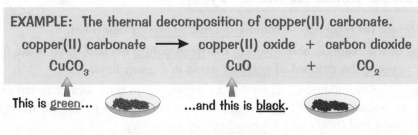

This is <u>green</u>...   ...and this is <u>black</u>.

It's too hot...
I'm breaking down...

This page is easy — let me break it down for you...

Here are three common types of <u>chemical reaction</u> to read up and <u>learn</u>. They're here because they're important, so make sure you commit them to memory ASAP. That way you'll have plenty of time to enjoy what's coming up on the next page — that's right, yet more fun and interesting chemical reactions.

More on Chemical Reactions

Chemical reactions always involve a transfer of energy to or from the surroundings.

In an Exothermic Reaction, Energy is Transferred Out

> An exothermic reaction is one which transfers energy to the surroundings.

1) Energy is usually given out by heating, so exothermic reactions involve a rise in temperature.

2) The best example of an exothermic reaction is combustion (see previous page). This gives out a lot of energy — it's very exothermic.

3) Many neutralisation reactions (page 53) and oxidation reactions (previous page) are exothermic.

4) Everyday uses of exothermic reactions include hand warmers.

In an Endothermic Reaction, Energy is Taken in

> An endothermic reaction is one where energy is taken in from the surroundings.

1) Energy is usually taken in by heating, so endothermic reactions involve a fall in temperature.

2) Endothermic reactions are much less common. Thermal decompositions (previous page) are a good example, since they involve a substance taking in energy and breaking down.

3) Everyday uses of endothermic reactions include sports injury packs. They take in energy and get very cold.

Catalysts Increase the Speed of a Reaction

> A catalyst is a substance which speeds up a chemical reaction, without being changed or used up in the reaction itself.

1) Catalysts come out of a reaction the same as when they went in — usually they just give the reacting particles somewhere to meet up and do the business. That means catalysts can be reused.

2) Chemical reactions need energy to get them started — usually through heating. Catalysts lower the minimum amount of energy needed for a reaction to happen.

3) This means a lower temperature can be used to carry out the reaction.

Catalysts Help Reduce Costs in Industry

1) Catalysts are very important for business — most industrial reactions use them.

2) By increasing the speed of the reaction and lowering the temperature needed, they make industrial reactions cheaper and increase the amount of product made in a given time.

3) There are some disadvantages to catalysts. They can be expensive to buy, and different reactions use different catalysts, so businesses can't get away with just buying one to use for everything. They also need to be cleaned and they can be 'poisoned' by impurities.

Catalysts are like my jokes — they can be used over and over

And they're not only used in industry... the enzymes in your body are a type of catalyst — without them, your chemical reactions would be too slow to keep you alive. There's more on enzymes on p. 8.

Balancing Equations

It's important to live a <u>balanced</u> life — that includes work, play, nutrition and <u>chemical equations</u>.

Chemical Equations *Show What Happens in a Reaction*

You can show what happens in a chemical reaction using:
1) A <u>WORD EQUATION</u> — where the <u>names</u> of the products and reactants are written out in <u>full</u>.
2) A <u>SYMBOL EQUATION</u> — which uses <u>chemical symbols</u> and <u>formulae</u> (see pages 35-37).
   A <u>balanced</u> symbol equation shows <u>how many</u> of each chemical react or are made in a reaction.

Chemical Equations *are Equal on Both Sides*

Oxygen gas is made up of pairs of atoms, called molecules — that's why it's O_2.

Here's an example of writing a balanced equation for burning magnesium in oxygen.

1) Write the <u>word equation</u>: magnesium + oxygen $\longrightarrow$ magnesium oxide

2) Write in the <u>chemical formulae</u> of all the reactants and products: $Mg + O_2 \longrightarrow MgO$

3) Check that the equation is <u>balanced</u> by <u>counting</u> the number of <u>each atom</u> on <u>both sides</u> of the equation. Then do steps A, B, C and D below to <u>balance</u> the atoms up one by one. Keep track of the <u>number</u> of atoms on <u>each side</u> as you go:

Left side of equation	Right side of equation
One Magnesium	One Magnesium
Two Oxygen	One Oxygen

A Find an element that doesn't balance and pencil in a number to try and sort it out.

There isn't enough <u>oxygen</u> on the <u>right side</u> of the equation — add "<u>2</u>" before MgO.

$$Mg + O_2 \longrightarrow 2MgO$$

B See where that gets you by counting up the atoms again.

Left side of equation	Right side of equation
One Magnesium	Two Magnesium
Two Oxygen	Two Oxygen

C Continue to chase the unbalanced atoms by going back to A) — pencil in a number before a formula, then see where it gets you when you count up the atoms.

There isn't enough <u>magnesium</u> on the <u>left side</u> of the equation — add a "<u>2</u>" before Mg.

$$2Mg + O_2 \longrightarrow 2MgO$$

D See where that gets you by counting up the atoms again.

Left side of equation	Right side of equation
Two Magnesium	Two Magnesium
Two Oxygen	Two Oxygen

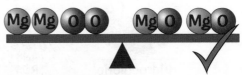

Done and dusted.

All things being equal — you'll be able to sort this out...

This is the <u>hardest</u> thing you'll ever be asked to do in this subject, so don't worry if you find it tricky.
<u>Learn the method</u> and then <u>try this one</u> out for size: $Na + Cl_2 \longrightarrow NaCl.$ *

Acids and Alkalis

The <u>pH scale</u> is what scientists use to describe how <u>acidic</u> or <u>alkaline</u> a substance is. <u>Universal indicator</u> takes on a colour based on the <u>pH</u> of the substance it's mixed with.

The pH Scale Shows the Strength of Acids and Alkalis

1) The <u>pH scale</u> goes from <u>0 to 14</u>.
2) Anything with a pH <u>below 7</u> is an <u>acid</u>. The <u>strongest</u> acid has <u>pH 0</u>.
3) Anything with a pH <u>above 7</u> is an <u>alkali</u>. The <u>strongest</u> alkali has <u>pH 14</u>.
4) A <u>neutral</u> substance has <u>pH 7</u> (e.g. water).

Indicators Are Special Dyes Which Change Colour

1) An indicator is just something that <u>changes colour</u> depending on whether it's in an <u>acid</u> or in an <u>alkali</u>.

2) <u>Litmus paper</u> is quite a popular indicator, but it only tells us whether a liquid is an <u>acid</u> or an <u>alkali</u> — it <u>does not</u> say how <u>strong</u> it is. <u>Acids</u> turn litmus paper <u>red</u> and <u>alkalis</u> turn it <u>blue</u>.

3) <u>Universal indicator</u> solution is a very useful <u>mixture of dyes</u> which gives the colours shown in a <u>pH chart</u>.

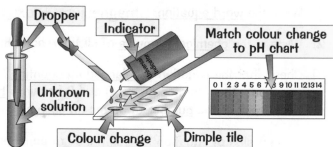

A pH Chart Shows How Strong an Acid or Alkali is

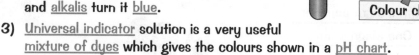

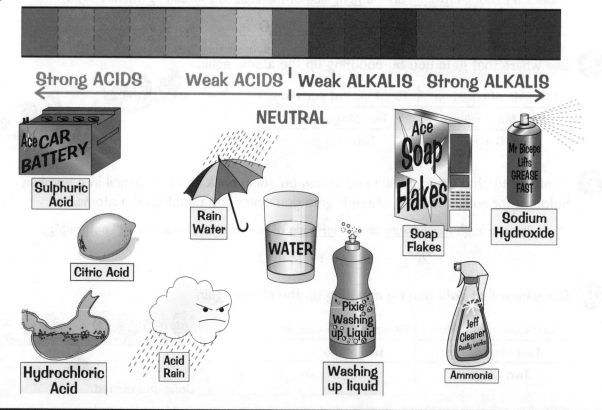

pHew — the end of another page...

All the pictures above are positioned <u>directly below</u> their actual <u>pH</u> on the pH scale. Make sure you know <u>where</u> each one of them goes, so you don't get your acid rain and your soap flakes mixed up.

Neutralisation Reactions

You might have done something like this in the lab and, if not, I bet you will pretty soon.
Make sure you <u>know</u> all this stuff — it's pretty easy and a <u>super-useful</u> thing to know about.

Acids and *Alkalis* Neutralise *Each Other*

1) <u>Acids</u> react with <u>alkalis</u> to form a <u>neutral</u> solution of a <u>salt</u> and <u>water</u>:

$$acid\ +\ alkali\ \longrightarrow\ salt\ +\ water$$

2) This is known as a <u>neutralisation</u> reaction because the products have a <u>neutral pH</u>, i.e. a pH of 7.

Making Salts *by Neutralisation*

Making <u>salts</u> is pretty easy — you just need a steady hand and a lot of time. A bit like whisking eggs.

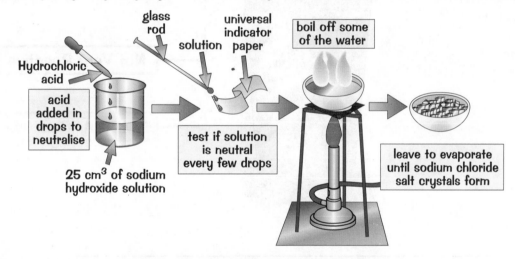

1) Wearing <u>eye protection</u>, add an <u>acid</u> to an <u>alkali</u> dropwise with a pipette.
2) After every few drops, <u>remove</u> a <u>small sample</u> to check if the <u>pH is neutral</u> (pH 7).
3) Keep <u>adding acid</u> until the solution is <u>neutral</u>.
4) When it's neutral the solution is put in an <u>evaporating dish</u> and about two thirds of it can be <u>boiled off</u> to make a <u>saturated solution</u> of the salt.
5) Leave this solution <u>overnight</u> for the rest of the water to evaporate and nice <u>big salt crystals</u> will form. The <u>slower</u> the <u>crystallisation</u>, the <u>bigger</u> the crystals.

A saturated salt solution can't have any more salt dissolved in it. See p.39.

To Change the *Salt,* You Must *Change* the *Acid*

1) The <u>salt</u> you get out of the <u>neutralisation</u> reaction above depends on the <u>acid</u> you use.
2) The clue is normally in the <u>name</u>:

<u>Hydrochloric acid</u> reacts to make <u>chloride salts</u>...   like <u>sodium chloride</u>.

<u>Sulfuric acid</u> reacts to make <u>sulfate salts</u>...   like <u>copper sulfate</u>.

<u>Nitric acid</u> reacts to make <u>nitrate salts</u>...   like <u>sodium nitrate</u>.

It's fun making salts — but I wouldn't put them on yer chips...

Make sure you <u>know</u> which kinds of <u>salts</u> you get from which kind of <u>acid</u>. I could sit here and tell you that if you don't learn this stuff it'll come up at the worst possible opportunity, and that without knowing the really simple stuff on this page you'll get absolutely nowhere... but that would just be rubbing salt into the wounds.

Reactivity Series and Metal Extraction

You need to know which metals are most reactive — and which are least reactive.

The Reactivity Series — How Well a Metal Reacts

The Reactivity Series lists metals in order of their reactivity towards other substances.

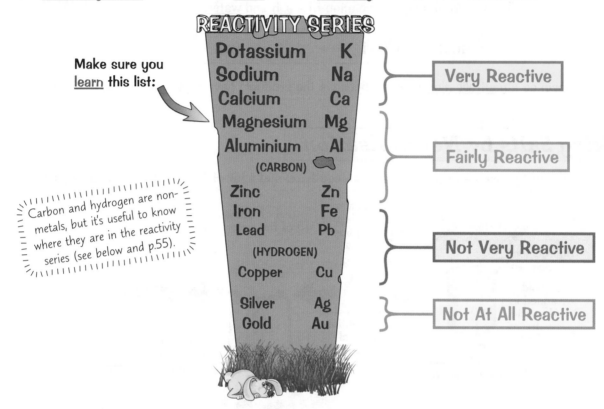

Make sure you learn this list:

REACTIVITY SERIES

Potassium	K	Very Reactive
Sodium	Na	
Calcium	Ca	
Magnesium	Mg	Fairly Reactive
Aluminium	Al	
(CARBON)		
Zinc	Zn	
Iron	Fe	Not Very Reactive
Lead	Pb	
(HYDROGEN)		
Copper	Cu	
Silver	Ag	Not At All Reactive
Gold	Au	

Carbon and hydrogen are non-metals, but it's useful to know where they are in the reactivity series (see below and p.55).

Some Metals Can Be Extracted With Carbon

1) Metals are usually mined as ores — rocks containing different metals and metal compounds (usually metal oxides — see page 56).

2) A metal can be extracted from its ore by reduction using carbon. When an ore is reduced, oxygen is removed from it. E.g. the oxygen is removed from iron oxide to extract the iron:

> iron oxide + carbon ⟶ iron + carbon dioxide

3) Only metals that are less reactive than carbon (i.e. metals below carbon in the reactivity series) can be extracted from their ore using carbon.

4) Metals that are more reactive than carbon need to be extracted using electrolysis (where electrical energy splits up the ore into the elements that make it up).

5) Some metals, like silver and gold, are pretty unreactive, so they're often found in their pure form.

Potassium
Sodium
Calcium
Magnesium
Aluminium
—CARBON—
Zinc
Iron
Lead
Copper
Silver
Gold

Metal extraction — sounds painful...

Let's face it, this is pretty basic stuff on metals. One of the first things you should know about metals is the difference in how reactive they are. In Chemistry, the reactivity of a metal is far and away the most important feature of it, because that's what decides how it will behave in every reaction it's faced with.

Reaction of Metals with Acids

One more page on metals to test your mettle — it's not so bad though, I promise. You don't need to know about each individual reaction, just how the reactivity of each metal affects it. Simple, no?

Reacting Metals With Dilute Acid

$$\text{metal} + \text{acid} \longrightarrow \text{salt} + \text{hydrogen}$$

All acids contain hydrogen — so the hydrogen here comes from the acid.

1) Metals above hydrogen in the reactivity series (see page 54) will react with acids to make a salt and hydrogen.
2) The metals below hydrogen in the reactivity series don't react with acids.
3) The reaction becomes less and less exciting as you go down the series.

More Reactive Metals React More Violently

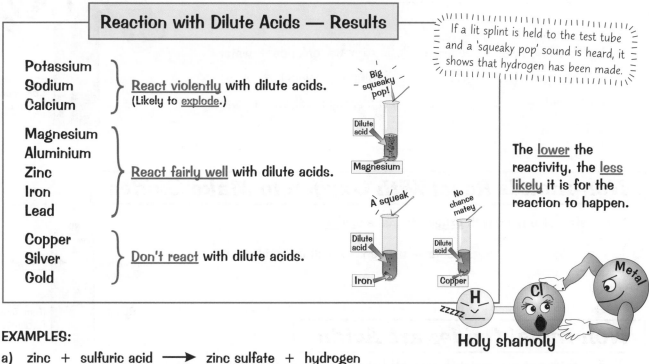

Reaction with Dilute Acids — Results

If a lit splint is held to the test tube and a 'squeaky pop' sound is heard, it shows that hydrogen has been made.

Potassium
Sodium
Calcium
} React violently with dilute acids. (Likely to explode.)

Magnesium
Aluminium
Zinc
Iron
Lead
} React fairly well with dilute acids.

Copper
Silver
Gold
} Don't react with dilute acids.

Big squeaky pop!

Dilute acid

Magnesium

A squeak

No chance matey

Dilute acid

Dilute acid

Iron

Copper

The lower the reactivity, the less likely it is for the reaction to happen.

Holy shamoly

EXAMPLES:

a) zinc + sulfuric acid $\longrightarrow$ zinc sulfate + hydrogen

$$Zn + H_2SO_4 \longrightarrow ZnSO_4 + H_2$$

The zinc takes the place of the hydrogen in the acid because it's more reactive than the hydrogen.

b) sodium + hydrochloric acid $\longrightarrow$ sodium chloride + hydrogen

$$2Na + 2HCl \longrightarrow 2NaCl + H_2$$

The sodium takes the place of the hydrogen in the acid — again because it's more reactive than the hydrogen.

You're probably bored of metals now — time to reactivate...

It might seem like there's loads going on here, but really it's just the same principle repeated over and over. All the metals have roughly the same reaction with acids, some are just more violent than others (and some don't happen at all). All you need to know is the order of violentness. Is that a word? Let's go with yes.

Reactions of Oxides with Acids

<u>Oxides</u> are pretty self-explanatory — they've got <u>oxygen</u> in them somewhere...

Metals React With Oxygen to Make Oxides

Metals react with <u>oxygen</u> to make <u>metal oxides</u>.

E.g. magnesium + oxygen → magnesium oxide.

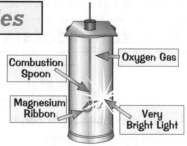

Metal Oxides are Alkaline

1) Metal oxides in solution have a <u>pH</u> which is <u>higher than 7</u> — i.e. they're <u>alkaline</u>.
2) So <u>metal oxides</u> react with <u>acids</u> to make a <u>salt</u> and <u>water</u>.

    acid + metal oxide ⟶ salt + water

EXAMPLES:
hydrochloric acid + copper oxide → copper chloride + **water**
sulfuric acid + zinc oxide → zinc sulfate + **water**
nitric acid + magnesium oxide → magnesium nitrate + **water**

pH more than 7

Non-metals React With Oxygen to Make Oxides

Non-metals also react with <u>oxygen</u> to make <u>oxides</u>.

E.g. sulfur + oxygen → sulfur dioxide.

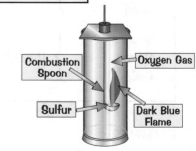

Non-metal Oxides are Acidic

1) The oxides of non-metals have a <u>pH below 7</u>. This means they're <u>acidic</u>.
2) So <u>non-metal oxides</u> will react with alkalis to make a <u>salt</u> and <u>water</u>.

    alkali + non-metal oxide ⟶ salt + water

EXAMPLE:
sodium hydroxide + silicon dioxide → sodium silicate + **water**

an alkali   a non-metal oxide

pH less than 7

Everyone has a good side, a bad side and an oxide...

There's not a lot on this page, which means you have no excuses whatsoever not to <u>learn</u> it. You might also notice some <u>similarities</u> between the reactions of <u>metals</u> and <u>non-metals</u> — replace "acid" with "alkali" and they're pretty much identical. With that said, you should probably try not to get the two confused.

Displacement Reactions

This page is pretty crammed, but the stuff on it is actually dead easy, I promise...

'Displacement' Means 'Taking the Place of'

A more reactive metal will displace a less reactive metal from its compound.

1) The reactivity series (see page 54) tells you which are the most reactive metals —
i.e. the ones which react most strongly with other things.

2) If you put a more reactive metal like magnesium into a solution of a less reactive metal compound, like copper sulfate, then magnesium will take the place of the copper — and make magnesium sulfate.

3) The "kicked out" metal then coats itself on the reactive metal, so we'd see copper.

4) This only happens if the metal added is more reactive — higher displaces lower. Got it?

A Reactivity Series Investigation

Method: Slap a bit of metal into some salt solutions and see what happens.

Results:

Magnesium	Magnesium	Iron	Zinc	Copper
1)	2)	3)	4)	5)
Deposit of copper	Deposit of zinc	Deposit of copper	Dull deposit of iron	No deposit

Salt Solution Used:
Copper Sulfate $CuSO_4$(aq)	Zinc Sulfate $ZnSO_4$(aq)	Copper Sulfate $CuSO_4$(aq)	Iron Sulfate $FeSO_4$(aq)	Zinc Sulfate $ZnSO_4$(aq)

Tube 1: The blue copper sulfate solution goes colourless and the copper coats the magnesium strip.
Magnesium must be more reactive than copper as it takes its place.

magnesium + copper sulfate ⟶ magnesium sulfate + copper

Tube 2: Zinc is seen coating the magnesium strip.
Magnesium must be more reactive than zinc as it takes its place.

magnesium + zinc sulfate ⟶ magnesium sulfate + zinc

Tube 3: The blue copper sulfate solution goes green and the copper coats the nail.
Iron must be more reactive than copper as it takes its place.

iron + copper sulfate ⟶ iron sulfate + copper

Tube 4: Iron is seen coating the zinc strip. Zinc must be more reactive than iron as it takes its place.

zinc + iron sulfate ⟶ zinc sulfate + iron

Tube 5: There's no reaction. Copper can't displace zinc — it's not reactive enough.

copper + zinc sulfate ⟶ no way

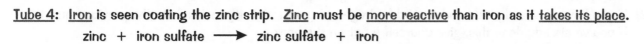

Most Reactive — Magnesium — Zinc — Iron — Copper — Least Reactive

Neutralisation is a Displacement Reaction

1) The hydrogen in hydrochloric acid is displaced (or replaced) by sodium from the sodium hydroxide (the alkali).

2) This makes NaCl and H_2O.

3) NaCl is sodium chloride — common salt.
And of course H_2O is water. Of course you knew.

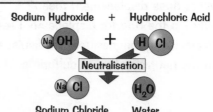

Sodium Hydroxide + Hydrochloric Acid

Na OH + H Cl

Neutralisation

Na Cl H_2O

Sodium Chloride   Water

Displace or datplace — wherever you do it, learn this stuff...

You've made it to the promised land — the end of another section. What every aspiring student dreams of.
Learn all the displacement stuff, then try not to misplace your knowledge when it comes to crunch time.

Section Summary

There's no use getting through a whole section of Chemistry if you can't summarise it with a handy set of questions that test everything you need to know. Luckily for you, that's exactly what this page is for. You must have heard it all before by now, and it's the usual shtick — work through the questions one by one, make sure you know everything, then maybe treat yourself to something sweet.

1) What happens to the atoms in a chemical reaction?

2) Does the mass change during a chemical reaction? Why or why not?

3) What's combustion?

4) What's the name of the process in which a chemical gains oxygen?

5) What's thermal decomposition?

6) What's formed when a metal carbonate breaks down by thermal decomposition?

7) What's the main difference between exothermic and endothermic reactions?

8)* If you put sodium in water, it catches fire and burns up. Is this reaction exothermic or endothermic?

9)* When ammonia breaks down to nitrogen and hydrogen, the temperature drops. Is this reaction exothermic or endothermic?

10) How does a catalyst affect the speed of a reaction?

11) Give two reasons why a chemical production plant might want to use catalysts.

12) Give two reasons why a chemical production plant might <u>not</u> want to use catalysts.

13)* Write a balanced symbol equation for: sulfur + oxygen $\longrightarrow$ sulfur dioxide (clue: $S + O_2 \longrightarrow$?)

14)* Write a balanced symbol equation for: calcium + oxygen $\longrightarrow$ calcium oxide

15) What pH does the strongest acid on a pH chart have? And the strongest alkali?

16) What pH does a neutral solution have?

17) What colour would universal indicator go if it was mixed with:
   a) a strong acid   b) a neutral solution   c) a strong alkali

18) What is neutralisation?

19) Outline the method to make common salt — sodium chloride.

20) Hydrochloric acid makes chloride salts — what salts does sulfuric acid make?

21) What kind of salts do you get from nitric acid?

22) List the reactivity series in the correct order. Take the first letter of each element and make up a rhyme to help you remember it — there, that'll cheer you up.

23) If you haven't already, add carbon and hydrogen to your reactivity series from the previous question. If you've already done this, give yourself a pat on the back.

24) Which metals in the reactivity series can be extracted from their ores using carbon? Which can't? Explain why they can't.

25) What do metals produce when they react with an acid?

26) Which metal will react the most violently with acid?

27) Are metal oxides in solution acidic, neutral or alkaline?

28) Give an example of a neutralisation reaction involving a metal oxide.

29) Are non-metal oxides in solution acidic, neutral or alkaline?

30) What does displacement mean?

31) What is the rule for displacement reactions?

32) Explain why magnesium can displace copper from copper sulfate.

33) In the neutralisation of sulfuric acid by potassium hydroxide, what displaces the hydrogen in the acid?

*Answers on page 108.

The Earth's Structure

Ever wondered what the planet's like on the inside? Well you're in for a treat with this page then.

The Earth Has a *Crust*, a *Mantle* and a *Core*

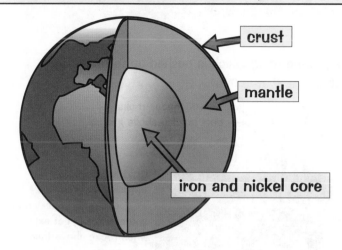

crust

mantle

iron and nickel core

The Earth is almost a sphere and it has a layered structure. A bit like a scotch egg. Or a peach.

1) We live on the crust — a thin, outer layer of solid rock.

2) Below that is the mantle.

3) The mantle is mostly solid, but deep down it can flow very slowly (like a liquid). This is because the temperature increases as you go deeper into the mantle.

4) At the centre of the Earth is the core. We think it's made of iron and nickel.

The Crust Contains *Minerals*

Elements and compounds make up minerals — and these make up rocks in the crust. E.g.

Elements	Compound	Mineral	Rock
Silicon & Oxygen	Silicon dioxide	Quartz	Granite

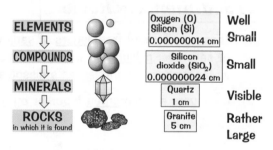

ELEMENTS ⇩
COMPOUNDS ⇩
MINERALS ⇩
ROCKS
in which it is found

Oxygen (O) Silicon (Si) 0.000000014 cm	Well Small
Silicon dioxide (SiO_2) 0.000000024 cm	Small
Quartz 1 cm	Visible
Granite 5 cm	Rather Large

The *Earth's Surface* is Made Up of *Tectonic Plates*

1) The crust and the upper part of the mantle are cracked into a number of large pieces. These pieces are called tectonic plates.

2) Tectonic plates are a bit like big rafts that 'float' on the mantle. They're able to move around.

3) The map shows the edges of the plates as they are now, and the directions they're moving in (red arrows).

4) Most of the plates are moving very slowly (a few centimetres a year).

5) Sometimes, the plates move very suddenly, causing an earthquake.

6) Volcanoes and earthquakes often happen where two tectonic plates meet.

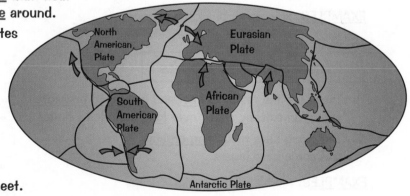

North American Plate

Eurasian Plate

South American Plate

African Plate

Antarctic Plate

Personally, I always cut off the crust...

You need to know the structure of Earth, i.e. what it would look like if you cut it open (which I wouldn't recommend) and what it's made of. That top diagram is your friend — learn it and learn it well. And, while we're on the subject, you'll need to learn all the words too. On the whole page. Phew.

Rock Types

Yep, there's <u>more than one</u> sort of rock. Who'd have thought it.

There are *Three Different* Types *of Rock*

1) Igneous **Rocks**

1) These are formed from <u>magma</u> (melted underground rock) which is pushed up to the surface of the crust — and often out through <u>volcanoes</u>.

2) They contain various minerals in randomly arranged <u>interlocking crystals</u>. The size of the crystals (or texture) depends on the speed of cooling. <u>Large</u> crystals mean that the rock has cooled <u>slowly</u>.

3) There are <u>two types</u> of igneous rocks: <u>extrusive</u> and <u>intrusive</u>.

<u>EXAMPLES</u>: basalt (extrusive),
granite (intrusive).

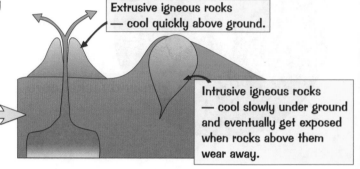

Extrusive igneous rocks — cool quickly above ground.

Intrusive igneous rocks — cool slowly under ground and eventually get exposed when rocks above them wear away.

2) *Sedimentary* **Rocks**

1) These are formed from <u>layers</u> of <u>sediment</u> (rock fragments or dead matter) laid down in lakes or seas over <u>millions</u> of years. Sedimentary rocks can also form when water evaporates and leaves a <u>dissolved solid</u> (like salt) behind.

2) The layers are <u>cemented</u> together by <u>other minerals</u>.

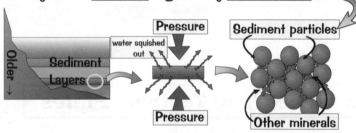

Older

Sediment Layers

water squished out

Pressure

Pressure

Sediment particles

Other minerals

OLDER

Fossils in the rock layers

3) <u>Fossils</u> can form in the sediments. These are the long dead <u>remains</u> of <u>plants</u> and <u>animals</u>. The <u>type</u> of fossil is used to work out the relative age of the rock.

<u>EXAMPLES</u>: limestone, chalk, sandstone.

3) *Metamorphic* **Rocks**

1) These are the result of <u>heat</u> and <u>increased pressure</u> acting on existing rocks over <u>long</u> periods of time.

2) They may have really <u>tiny crystals</u> and some have layers.

<u>EXAMPLES</u>: marble, slate, schist.

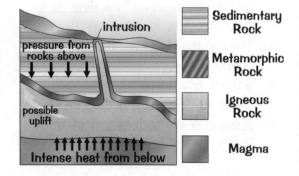

intrusion

pressure from rocks above

possible uplift

Intense heat from below

Sedimentary Rock

Metamorphic Rock

Igneous Rock

Magma

OK, sure — but doesn't Glam-Rock deserve a mention...

Three types of rock for you to know about — make sure you do. I agree they've got pretty scary-looking names but once you get over that little hurdle and just <u>learn them</u>, the rest all follows a lot easier. <u>Learn</u> the headings, then do a <u>mini-essay</u> for each one <u>from memory</u>. Draw out the <u>diagrams</u> too.

The Rock Cycle

The rock cycle involves changes to rocks both <u>inside</u> and <u>outside</u> the Earth.

The Rock Cycle Takes Millions of Years to Complete

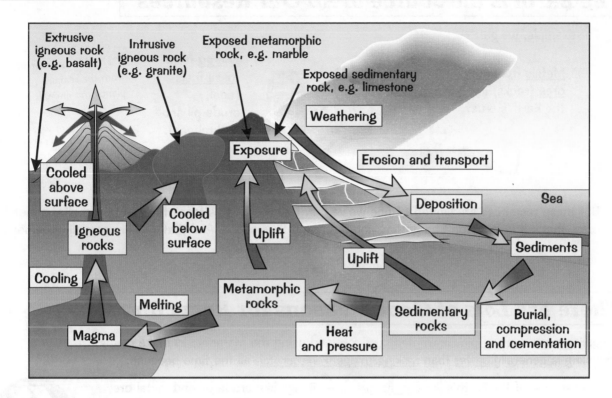

The rock cycle involves <u>changing</u> the three types of rock (<u>igneous</u>, <u>sedimentary</u> and <u>metamorphic</u>, see previous page) from one to another. This happens by:

1) <u>WEATHERING</u>: <u>breaking down</u> rocks into <u>smaller bits</u>.
   There are a few different ways this can happen, e.g.

2) <u>EROSION</u>: wearing down rocks, e.g. by rain.

3) <u>TRANSPORTATION</u>: moving the eroded bits of rock round the world by wind and water (mostly).

4) <u>DEPOSITION</u>: laying down of sediment.

5) <u>BURIAL/COMPRESSION/CEMENTATION</u>: squeezing and compressing the layers — eventually they form <u>sedimentary rocks</u>.

6) <u>HEAT/PRESSURE</u>: further squashing and heating — turns the rocks into <u>metamorphic rocks</u>.

> <u>Onion skin weathering</u> — this happens when the Sun <u>warms</u> the <u>surface</u> of a rock by <u>day</u> and by <u>night</u> it <u>cools</u> down. This causes the surface to <u>expand</u> and <u>contract</u>, and eventually it <u>breaks away</u>, like <u>peeling an onion</u>.
>
> <u>Freeze-thaw weathering</u> — when water <u>freezes</u>, it <u>expands</u>. If this happens in a <u>crack</u> in a rock it can make the crack <u>bigger</u>. After freezing and thawing many times, <u>bits break off</u>.

7) <u>MELTING</u>: intense heating makes the rock partially melt — that changes it to magma.

8) <u>COOLING</u>: solidification of the molten (melted) rock to form <u>igneous rocks</u>.

9) <u>EXPOSURE</u>: back to weathering and erosion again. Simple huh.
   (The <u>amount</u> of rock on the surface is always <u>about the same</u>, even though it's <u>weathered</u> away.)

The Rock Cycle's a bit like homework — it takes forever...

<u>Nine stages</u> of the rock cycle to learn there — make sure you know <u>what happens</u> at each stage, but also how each of the stages are <u>linked</u>. You'll really impress teach' if you can <u>explain</u> how a <u>sedimentary rock</u> changes into a <u>metamorphic rock</u> and so on. Learn it well and you'll be a rock (cycle) star before long.

Recycling

My mum is pretty big on recycling. You can't throw anything in the bin in our house. Turns out we're doing our bit to <u>save limited resources</u> though, so that's OK.

The Earth is the Source of All Our Resources

1) For example, we get:

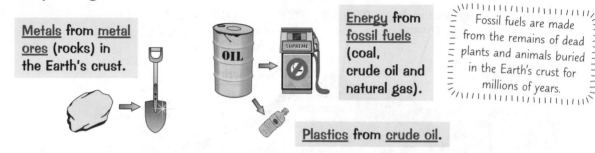

<u>Metals</u> from <u>metal ores</u> (rocks) in the Earth's crust.

<u>Energy</u> from <u>fossil fuels</u> (coal, crude oil and natural gas).

Fossil fuels are made from the remains of dead plants and animals buried in the Earth's crust for millions of years.

<u>Plastics</u> from <u>crude oil</u>.

2) But these resources are <u>limited</u>. Once we've <u>burnt</u> all the Earth's fossil fuels or <u>mined</u> all the metal ores, <u>that's it</u> — we <u>won't</u> be <u>getting any more</u> any time soon. And that's where <u>recycling</u> comes in.

There are Lots of Good Reasons for Recycling

Recycling means taking <u>old</u>, <u>unwanted products</u> and using the <u>materials</u> to make <u>new stuff</u>. Recycling is generally <u>better</u> than <u>making things from scratch</u> all the time because:

1) It uses <u>less</u> of the Earth's <u>limited resources</u> — things like crude oil and metal ores.
2) It uses <u>less energy</u> — which usually comes from burning fossil fuels.
3) Energy is expensive — so recycling tends to <u>save money</u> too.
4) It makes <u>less rubbish</u> — which would usually end up in <u>landfill sites</u> (<u>rubbish dumps</u>).

> <u>Example — recycling aluminium cans:</u>
>
> 1) If aluminium <u>wasn't recycled</u>, more <u>aluminium ore</u> would have to be <u>mined</u>.
> 2) Mining costs <u>money</u> and uses loads of <u>energy</u>. It also makes a <u>mess</u> of the <u>landscape</u>.
> 3) The ore then needs to be <u>transported</u> and the aluminium <u>extracted</u> — which uses <u>more energy</u>.
> 4) It then <u>costs</u> to send the <u>used aluminium</u> to <u>landfill</u>.
>
> It's a complex calculation, but for every <u>1 kg</u> of aluminium cans that are recycled, you <u>save</u>:
> - <u>95%</u> of the <u>energy</u> needed to mine and extract 'fresh' aluminium,
> - <u>4 kg</u> of aluminium ore,
> - a <u>lot</u> of waste.
>
>  It's really efficient to recycle aluminium.

It's usually <u>more efficient</u> (in terms of energy and cost) to <u>recycle materials</u> rather than throw them away and produce new ones. But the efficiency <u>varies</u> depending on what it is you're recycling. E.g. you get an energy saving of <u>95%</u> by recycling <u>aluminium</u>, but less with <u>plastics</u> (<u>70%</u>) and <u>steel</u> (<u>60%</u>).

Recycle this book — but wait till you've finished KS3 science

Not all materials are as efficient to recycle as aluminium. But even if the energy and cost savings are relatively small, you could still be <u>saving</u> precious <u>limited resources</u> and creating <u>less waste</u> — which can only be a good thing. There's nothing too tricky on this page, so <u>get it all learnt</u>. Pronto.

The Carbon Cycle

Carbon is a very important element because it's part of all living things.
As shown below, it's constantly recycled through the environment.

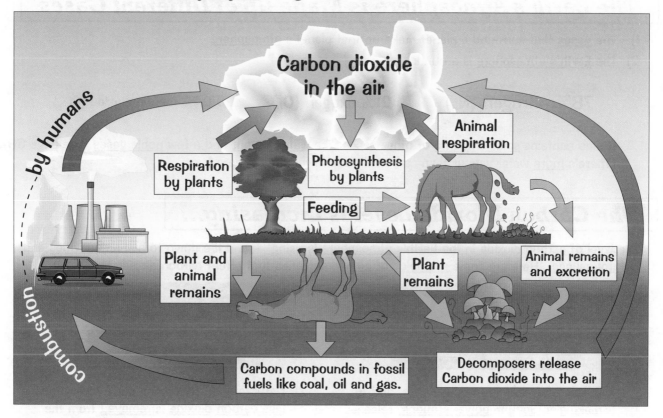

Learn these points:

1) Photosynthesis Removes Carbon Dioxide from the Air

1) Green plants and algae take in carbon dioxide from the air during photosynthesis (see p.19).

2) The plants and algae use the carbon to make carbohydrates, fats and proteins.

2) Carbon is Passed Along the Food Chain When Animals Feed

1) Some of the carbon in plants is passed on to animals when they eat the plants.

2) The animals then use the carbon to make fats and proteins of their own. The carbon moves along the food chain when the animals are eaten by other animals.

3) Respiration and Combustion Return Carbon Dioxide to the Air

1) Some carbon is returned to the air as carbon dioxide when plants and animals respire (see p.4).

2) When plants and animals die, decomposers (like bacteria and fungi) feed on them. Decomposers also feed on animal waste. When the decomposers respire, carbon dioxide is returned to the air.

3) Some dead plant and animal remains get buried and eventually form fossil fuels. When fossil fuels are burnt (combustion) this releases carbon dioxide back into the air.

Carbon Cycle — you know, like Bradley Wiggins'...

Another cycle for you to learn here. Start by learning the headings — they're the key to how the whole cycle works. Then focus on the details. And remember, photosynthesis removes carbon (in the form of carbon dioxide) from the air. Respiration and combustion put it back again. Well I'm glad we cleared that up.

The Atmosphere and Climate

It's important to know exactly what you're <u>breathing</u> in and out. So read this page and find out.

The Earth's Atmosphere *is Made Up of* Different Gases

1) The <u>gases</u> that surround a planet make up that planet's <u>atmosphere</u>.
2) The <u>Earth's atmosphere</u> is around:

| 78% nitrogen (N_2) | 21% oxygen (O_2) | 0.04% carbon dioxide (CO_2) |

It also contains <u>small amounts</u> of other gases, like <u>water vapour</u> and a few <u>noble gases</u> (see page 36).
(There's <u>more</u> water vapour than carbon dioxide in the atmosphere.)

The Carbon Dioxide Level *is Increasing...*

The level of carbon dioxide in the Earth's <u>atmosphere</u> is rising — and it's down to <u>human activities</u> and <u>natural causes</u>. Here are some <u>examples</u> of human activities that affect carbon dioxide levels:

1) <u>Burning fossil fuels</u> to power <u>cars</u>, and to make <u>electricity</u> in <u>power stations</u>, releases lots of carbon dioxide into the atmosphere.

2) <u>Deforestation</u> (chopping down trees) means <u>less carbon dioxide</u> is <u>removed</u> from the atmosphere by <u>photosynthesis</u>.

...Which is Affecting the Earth's Climate

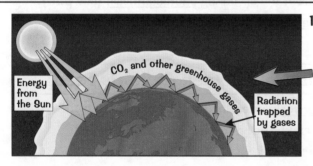

1) Carbon dioxide is what's known as a <u>greenhouse gas</u>. This means it <u>traps energy</u> from the <u>Sun</u> in the <u>Earth's</u> <u>atmosphere</u>. This <u>stops</u> a lot of energy from <u>being lost</u> into <u>space</u> and helps to keep the <u>Earth warm</u>.

This is a bit like what happens in a greenhouse. The Sun shines in, and the glass helps keep some of the energy in.

2) But the <u>level</u> of <u>carbon dioxide</u> (and a few <u>other greenhouse gases</u>) is <u>increasing</u>. The long term <u>trend</u> also shows that the <u>temperature</u> of Earth is <u>increasing</u>. Some scientists believe this is <u>due to</u> the rise in carbon dioxide levels.

3) This increase in the Earth's temperature is called <u>global warming</u>.

4) Global warming is a type of <u>climate change</u>. It could have some <u>serious effects</u>, e.g.

• <u>Glaciers</u> and <u>ice sheets</u> covering Greenland and Antarctica may <u>melt faster</u>, which could cause <u>sea levels</u> to <u>rise</u> and coastal areas to <u>flood</u>.

• <u>Rainfall patterns</u> could change, which might make it <u>harder</u> for farmers to <u>grow crops</u>.

This page is so atmospheric...

You need to know the <u>composition</u> of the Earth's atmosphere — so get learning those <u>percentages</u> at the top of the page. You also need to understand how <u>human activities</u> are affecting the <u>Earth's climate</u> — it's important for both <u>KS3 Science</u> and the <u>future of the planet</u>. So there's no excuse not to know the details.

Section Summary

Well there we are. The end of Section 7. All you have to do now is learn it all. And yes you've guessed it, here below are some lovely questions I prepared earlier. It's no good just idly going through them and managing half-baked answers to one or two that take your fancy. Make sure you can answer all of them.

1) The Earth is covered with a thin outer layer of rock. What is this layer called?

2) What is the name of the structure between the outer layer of rock and the Earth's core? Explain how this structure is both a solid and a liquid.

3) Which two metals do we think the Earth's core is made of?

4) Name a mineral present in the Earth's crust. Say what elements it contains.

5) What are tectonic plates?

6) How are igneous rocks formed?

7) What determines the size of the crystals in igneous rock?

8) How do sedimentary rocks form?

9) The dead remains of plants and animals can become trapped in sedimentary rocks. What are these remains called?

10) How do metamorphic rocks form?

11) Give two examples of: a) igneous rocks, b) sedimentary rocks, c) metamorphic rocks.

12) Draw out the full diagram of the rock cycle with all the labels.

13) What must happen to sedimentary rocks to turn them into metamorphic rocks?

14) What must happen to metamorphic rocks to turn them into igneous rocks?

15) Name two limited resources we get from the Earth.

16) Give four reasons why it's important to recycle materials.

17) How is carbon dioxide removed from the air by plants?

18) How does carbon get from the air into your body?

19) How do plants, animals and decomposers all return carbon dioxide to the air?

20) How else is carbon dioxide returned to the air?

21) What percentage of the Earth's atmosphere is: a) nitrogen, b) oxygen, c) carbon dioxide?

22) Name one other gas present in the Earth's atmosphere.

23) Give two human activities that are increasing the level of carbon dioxide in the atmosphere. Say why each one has an effect on the level of CO_2.

24) How does carbon dioxide help to keep the Earth warm?

25) What is global warming? What's causing it?

26) Describe two possible effects of global warming.

Energy Transfer

Ah, energy transfer. Everything you do involves <u>energy transfer</u>, which makes this a pretty important page.

Energy **Can Be** *Stored*

There are <u>seven</u> stores of <u>energy</u>. Here are some <u>examples</u> of each type:

Thermal Energy Store

<u>Everything</u> has some <u>energy</u> in its thermal energy store — the <u>hotter</u> it is, the <u>higher</u> its <u>temperature</u> and the <u>more</u> energy is in its thermal energy store.

Gravitational Potential Energy Store

Anything in a <u>gravitational field</u> (i.e. anything that can <u>fall</u>) has energy in its <u>potential energy store</u> — the <u>higher</u> it goes, the <u>more</u> it has.

Magnetic Energy Store

Two <u>magnets</u> that <u>attract</u> or <u>repel</u> each other have <u>energy</u> in their <u>magnetic energy stores</u>.

There's more on magnets on page 99.

Kinetic (Movement) Energy Store

Anything that <u>moves</u> has <u>energy</u> in its <u>kinetic energy store</u>.

Chemical Energy Store

Anything with <u>energy</u> which can be released by a <u>chemical reaction</u> — things like food, fuels and batteries.

Elastic Potential Energy Store

Anything <u>stretched</u> has <u>energy</u> in its <u>elastic energy store</u> — things like rubber bands, springs, knickers, etc.

Electrostatic Energy Store

Two <u>electric charges</u> that <u>attract</u> or <u>repel</u> each other have <u>energy</u> in their <u>electrostatic energy stores</u>.

 positive charge   negative charge

See page 98 for more.

Energy **Can Be** *Transferred* **Between Stores**

Whenever (pretty much) anything happens to an object, <u>energy</u> is <u>transferred</u> from one store to another — the store of energy you transfer to <u>increases</u> and the store of energy you transfer from <u>decreases</u>.

When you <u>drop</u> an object, it <u>moves</u> through a <u>gravitational field</u>. This causes energy to transfer <u>from</u> its <u>gravitational energy</u> store to its <u>kinetic energy</u> store.

When you <u>burn</u> fuel, energy is transferred from the fuel's store of <u>chemical energy</u> to the <u>thermal energy</u> store of the surroundings.

When you switch on this <u>electrical circuit</u>, energy is transferred from the <u>chemical energy</u> store in the <u>battery</u> to the <u>kinetic energy</u> store of the <u>motor</u>. As the motor <u>turns</u>, parts of it <u>rub</u> together — this causes some energy to be transferred from the <u>kinetic energy</u> store to the <u>thermal energy</u> store.

A <u>stretched object</u>, like a <u>spring</u>, has an <u>elastic energy</u> store. When it's released, the energy in the <u>elastic energy</u> store <u>decreases</u> quickly as it is transferred to the <u>kinetic energy</u> store.

<u>Food</u> has energy stored in <u>chemical energy</u> stores. When you eat food, it is <u>metabolised</u> (changed during chemical processes inside your body), which <u>releases</u> (transfers) the energy in the food. You can then use the energy for <u>useful things</u> like walking, keeping warm and studying science.

Energy Transfer

There are Four Ways of Transferring Energy

The four main ways you can transfer energy between stores are:

Mechanically

When a <u>force</u> makes something <u>move</u> (see page 67). E.g. if an object is <u>pushed</u>, <u>pulled</u>, <u>stretched</u> or <u>squashed</u>.

By Heating

When energy is transferred from <u>hotter</u> objects to <u>colder</u> objects (see page 68).

Electrically

When <u>electric charges</u> move around an electric <u>circuit</u> due to a potential difference (see page 95).

By Light and Sound

When <u>light</u> or <u>sound</u> waves (see Section 10) carry energy from <u>one place</u> to <u>another</u>.

Energy is Transferred When a Force Moves an Object

When a <u>force moves</u> an object through a <u>distance</u>, <u>energy is transferred</u>.

Energy transferred is the same as work done — see page 83.

1) Whenever something <u>moves</u>, something else is supplying some sort of '<u>effort</u>' to move it.

2) The thing putting in the <u>effort</u> needs a <u>supply</u> of <u>energy</u> (from <u>fuel</u> or <u>food</u>, etc.).

3) It then <u>transfers energy</u> by <u>moving</u> the object — the supply of energy is transferred to <u>kinetic energy stores</u>.

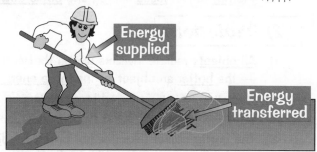

Energy supplied

Energy transferred

Energy Transferred, Force and Distance are Linked

1) To find how much <u>energy</u> has been <u>transferred</u> (in joules), you just multiply the <u>force in N</u> by the <u>distance moved in m</u>.

> Energy Transferred (in joules, J) = Force (in newtons, N) × Distance (in metres, m)

EXAMPLE: Some farmers drag an old tractor tyre <u>5 m</u> over rough ground. They pull with a total force of <u>340 N</u>. Find the energy transferred.

ANSWER: Energy transferred = force × distance
= <u>340 × 5 = 1700 J</u>

2) So, if a machine transfers a <u>certain amount</u> of energy, the amount of <u>force</u> it can apply and the <u>distance</u> over which it can apply it are <u>linked</u> — if one goes up, the other must come down.

3) So the machine can apply a <u>large force</u> over a <u>small distance</u>, or a <u>small force</u> over a <u>large distance</u>.

Start of a new section — I'm feeling energised...

There's plenty to learn here. Make sure you know how <u>energy</u> can be <u>stored</u> and <u>transferred</u> (learn all the <u>examples</u> on the previous page). You also need to know that <u>equation</u> to work out how much energy's transferred when a force moves an object. Cover up the pages and scribble down everything you remember.

Energy Transfer by Heating

Energy can be transferred between objects by heating.

Energy is Transferred From Hot Objects to Cooler Ones

1) When there's a temperature difference between two objects, energy will be transferred from the hotter one to the cooler one (so the hotter object will cool down and the cooler object will heat up).

2) This carries on until the objects reach thermal equilibrium — the point at which they're both the same temperature.

   You need to know about two ways in which energy can be transferred between objects by heating:

1) Conduction

1) When an object is heated, the particles in the object start vibrating more — they gain energy in their kinetic energy stores.

2) Conduction occurs when vibrating particles pass on their extra energy to neighbouring particles.

3) It only happens when particles can bump into each other, so the objects must be touching.

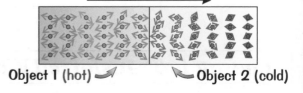

Energy transfer

Object 1 (hot)   Object 2 (cold)

4) Particles in the hotter object vibrate faster than particles in the cooler object. When the particles in the hot object bump into the particles in the cold object, energy is transferred. This means the hot object loses energy and cools down and the cold object gains energy and heats up.

2) Radiation

1) All objects radiate invisible waves that carry energy to the surroundings — the hotter an object is, the more energy it radiates.

2) Radiation isn't transferred by particles, so the objects don't need to be touching.

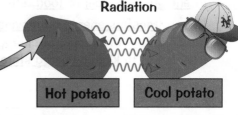

Radiation

Hot potato   Cool potato

3) The hotter object (like this hot potato) radiates more energy than the cooler object. The hotter object radiates more energy than it absorbs, so it cools down.

4) The cooler object absorbs some of the radiation from the hot object. It absorbs more energy than it radiates, so it heats up.

Insulators Can Slow Down the Rate of Energy Transfer

1) Some materials transfer energy more quickly than others. Objects made from conductors (e.g. metals) will transfer energy more quickly than objects made from insulators (e.g. plastics).

2) Wrapping an object in an insulator will slow down the rate at which it transfers energy to and from surrounding objects. So insulators help keep hot objects hot, and cold objects cold.

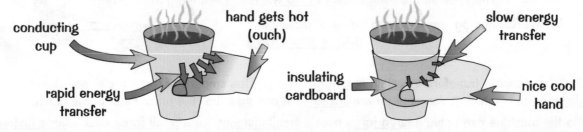

conducting cup — hand gets hot (ouch) — slow energy transfer — insulating cardboard — nice cool hand — rapid energy transfer

*I love energy transfer — it's totally rad*iation...

Ever wondered why metals at room temperature feel cold when you touch them? Metals are really good conductors, which means they draw energy away from your hand really quickly. Insulators like plastics and wood feel warmer even though they're the same temperature because they don't conduct energy as well.

Conservation of Energy

It's not that energy <u>can</u> be transferred, but more that it <u>has</u> to be. Not necessarily in a <u>useful</u> way, mind.

The Principles of Conservation of Energy

Scientists have only been studying energy for about two or three hundred years so far, and in that short space of time they've already come up with two "<u>Pretty Important Principles</u>" relating to energy. <u>Learn</u> them <u>really well</u>:

THE PRINCIPLE OF CONSERVATION OF ENERGY

Energy can never be CREATED nor DESTROYED — it's only ever TRANSFERRED from one store to another.

That means energy never simply <u>disappears</u> — it always <u>transfers</u> to another store. This is another <u>very useful principle</u>:

Energy is ONLY USEFUL when it's TRANSFERRED from one store to another.

Most Energy Transfers are Not Perfect

1) Useful devices are <u>useful</u> because they <u>transfer energy</u> from <u>one store</u> to <u>another</u>.

2) <u>Some energy</u> is always <u>lost</u> in some way, nearly always by <u>heating</u>.

3) As the diagram shows, the <u>energy input</u> will always end up coming out partly as <u>useful energy</u> and partly as <u>wasted energy</u> — but <u>no energy is destroyed</u>:

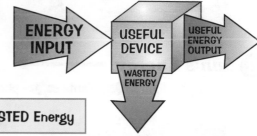

Total Energy INPUT = The USEFUL Energy + The WASTED Energy

You Can Also Draw Energy Transfer Diagrams

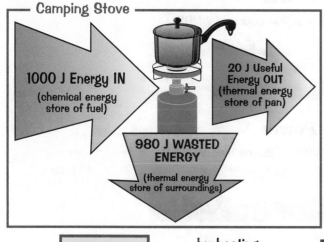

— Camping Stove —

1000 J Energy IN
(chemical energy store of fuel)

20 J Useful Energy OUT
(thermal energy store of pan)

980 J WASTED ENERGY
(thermal energy store of surroundings)

1) You can show how <u>energy</u> moves between <u>stores</u> by drawing an <u>energy transfer diagram</u> (see below).

2) <u>Rectangles</u> are used to represent the different <u>stores</u>.

3) Draw an <u>arrow</u> to show energy being <u>transferred</u> and <u>label</u> it with the <u>method</u> of transfer.

4) If there's <u>more than one</u> transfer, draw an arrow for each one, each going to a different store.

| CHEMICAL ENERGY STORE OF FUEL | by heating → | THERMAL ENERGY STORE OF PAN |
| | by heating → | THERMAL ENERGY STORE OF SURROUNDINGS |

No mum I'm not slacking — I'm just conserving energy...

This stuff's super <u>important</u>, so make sure you can wrap your head around it before ploughing on. Remember, energy's only ever <u>transferred</u> to other stores. No matter how hard you try, you <u>can't</u> ever create or destroy energy. It's just <u>impossible</u>. Futile. Unattainable. 100% completely out of the question.

Energy Resources

The <u>Sun</u>'s a useful little critter. It provides us with oodles of <u>energy</u> and asks for nothing in return.

The Sun is the Source of Our Energy Resources

Most of the <u>energy</u> around us <u>originates</u> from the <u>Sun</u>. The Sun's <u>energy</u> is really useful for supplying our energy demands. Often the Sun's energy is <u>transferred</u> to <u>different stores</u> before we use it.

Learn These Six Energy Transfer Chains

1 *Sun's Energy* ⟶ *Coal, Oil, and Gas (Fossil Fuels)*

Sun ⟹ light ⟹ photosynthesis ⟹ dead plants/animals ⟹ FOSSIL FUELS

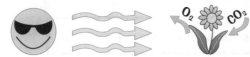

2 *Sun's Energy* ⟶ *Biomass (e.g. Wood)*

Sun ⟹ light ⟹ plants ⟹ photosynthesis ⟹ BIOMASS (wood)

3 *Sun's Energy* ⟶ *Food*

Sun ⟹ light ⟹ plants ⟹ photosynthesis ⟹ BIOMASS (food)

O_2 ... CO_2

4 *Sun's Energy* ⟶ *Wind Power*

Sun ⟹ heats atmosphere ⟹ air circulates ⟹ causes WINDS

Land warm   Sea cool

We can use the energy in the kinetic energy stores of the wind and waves to turn turbines and generators, giving us electricity.

5 *Sun's Energy* ⟶ *Wave Power*

Sun ⟹ heats atmosphere ⟹ causes WINDS ⟹ causes WAVES

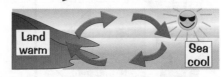

6 *Sun's Energy* ⟶ *Solar Cells*

Sun ⟹ light hits solar cells ⟹ generates ELECTRICITY

Baby you light up my world like nobody else...

...the way you give energy gets me overwhelmed. The <u>Sun</u>'s pretty amazing — without it we'd be totally scuppered. Don't take it for granted, because it supplies pretty much <u>all the energy</u> we have here on Earth.

Generating Electricity

We can use the energy we get from the Sun to generate electricity, in lots of different ways...

There Are Different Ways of Generating Electricity

1) There are a variety of different fuels that people use in their homes, e.g. coal is used for fires, gas is used for cookers, etc. But most homes these days rely on electricity for most of their energy needs.

2) We can use energy resources (see previous page) to generate electricity.

3) At the moment we generate most of our electricity by burning fossil fuels.

4) Most ways of generating electricity turn a turbine and a generator — the generator transfers the energy in kinetic energy stores electrically.

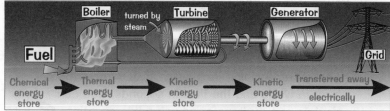

5) Energy resources that we use to generate electricity can be split into two groups — non-renewable and renewable.

Non-renewable Energy Resources Will Run Out

1) Fossil fuels took millions of years to come about — and only take minutes to burn.

2) Once they've been taken from the Earth — that's it, they're gone, (unless you're gonna wait around a few more million years for more to be made).

3) There'll come a time when we can't find any more and then we could have a problem.

4) We need to reduce the amount of fossil fuels we use, so they won't run out as quickly. The answer is:

> i) Save energy (e.g. turn lights off, drive cars with more fuel-efficient engines).
> ii) Recycle more (see page 62).
> iii) Use more renewable energy resources (see below).

Renewable Energy Resources Won't Run Out

As long as the Sun still shines...

1 The WIND will always blow
 — and turn turbines to generate electricity.

2 PLANTS will always grow
 — which can be burnt to generate electricity.

3 WAVES will always be made
 — and drive generators to make electricity.

4 SOLAR cells will always work
 — and use light to make electricity.

One of the drawbacks of renewable energy is that it's not always available like non-renewable resources. For example, there will always be wind, but it isn't always blowing with the same strength, solar cells are only useful during the day etc.

Think on pal — this'll affect all your generation...

Don't call renewable energy resources "re-usable" — no no no no no. They'll renew themselves, like trees will grow again if replanted, etc. But once a tree is burnt you can't re-use that particular tree. Don't let the name put you off though — renewable resources are, on the whole, pretty darn great.

The Cost of Electricity

Electricity ain't free you know — ask your mum and dad. At least the cost is pretty easy to calculate.

You Can Calculate the Energy an Appliance Transfers

1) Anything that needs electricity to work is an electrical appliance — electricity is its "fuel".

2) All electrical appliances transfer energy electrically to stores of energy.
Energy can be measured in joules (J), kilojoules (kJ) or kilowatt-hours (kWh).

3) Power tells you how fast something transfers energy. It's usually measured in watts (W)
or kilowatts (kW). 1 kW = 1000 W.

4) The total amount of energy transferred
depends on the amount of time something's
switched on for, and its power. If you know the
power in watts and the time in seconds, you can
calculate energy transferred using this equation:

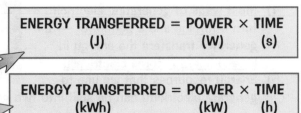

ENERGY TRANSFERRED = POWER × TIME
(J)   (W)   (s)

5) If you know the power in kilowatts and
the time in hours, you can use this equation:

ENERGY TRANSFERRED = POWER × TIME
(kWh)   (kW)   (h)

Electricity Meters Record How Much Electricity is Used

Electricity meters record the amount of energy transferred in kWh. You can use them
to work out the energy transferred over different periods of time, e.g. at day and at night:

EXAMPLE: Ganesh wants to find out how much electricity he uses during
the day compared to during the night. He writes down his
meter reading at three different times during a 24 hour period:
6pm = 44281.25 kWh   6am = 44284.76 kWh   6pm = 44296.12 kWh
Does he use more electricity during the day or during the night?

`44281.25 kWh`

ANSWER: Energy from 6pm to 6am (i.e. during the night) = 44284.76 – 44281.25 = 3.51 kWh
Energy from 6am to 6pm (i.e. during the day) = 44296.12 – 44284.76 = 11.36 kWh
So he uses more electricity during the day.

Calculating the Cost of Electricity

Domestic fuel bills charge by the kilowatt-hour.
You can calculate what your electricity bill should be with this handy little formula:

COST = Energy transferred (kWh) × PRICE per kWh   Cost = kWh × Price

EXAMPLE: Electricity costs 16p per kWh. At the start of last month, Jo's electricity meter
reading was 42729.66 kWh. At the end of the month it was 43044.91 kWh.
Calculate the cost of her electricity bill last month.

ANSWER: Energy transferred = 43044.91 – 42729.66 = 315.25 kWh
Cost = Energy transferred × price = 315.25 × 16 = 5044p = £50.44

Many homes use gas as a fuel, e.g. for gas central heating, gas cookers etc.
Your gas bill is calculated using the energy used in kWh, just like your electricity bill.

Feeling left in the dark? Maybe you didn't pay your bills...

Ever wondered what those little numbers on your electricity meter read? No? Oh, I guess it's just me then.
It's good to check the power company's charged you the right amount every now and then — and now you
can do just that with a nice little formula. Lucky you. Make sure you know how to use it — it's dead useful.

Comparing Power Ratings and Energy Values

If you want to know how much energy an <u>appliance</u> uses you can work it out using its <u>power rating</u>. And if you want to know how much energy is in your <u>food</u>, just look on the label. This energy stuff's everywhere.

Power Ratings *of Appliances*

1) The power rating of an appliance is the <u>energy</u> that it uses <u>per second</u> when it's operating at its recommended maximum power (i.e. when it's plugged into the mains).

2) You can <u>work out</u> the energy transferred by an appliance in a certain <u>time</u> if you know its <u>power rating</u>. To do this you need to use the <u>equations</u> on the <u>previous page</u>.

The Energy Transferred **Depends on the** Power Rating

The higher the <u>power rating</u> of an appliance, the <u>more energy</u> it transfers in a <u>given amount of time</u>. You can compare how much energy is transferred by appliances with <u>different power ratings</u>.

EXAMPLE: How much energy is transferred by a 1.5 kW electric heater compared to a 4 kW electric heater, when they're both left on for 1.5 hours?

ANSWER: <u>Energy transferred (kWh) = power rating (kW) × time (h)</u>.
Energy transferred by the 1.5 kW heater = 1.5 × 1.5 = 2.25 kWh.
Energy transferred by the 4 kW heater = 4 × 1.5 = 6 kWh.

So the 4 kW heater transfers (6 − 2.25) <u>3.75 kWh more energy</u> than the 1.5 kW heater in 1.5 hours.

Remember, transferring energy <u>costs money</u>. So an appliance with a <u>higher power rating</u> will cost <u>more to run</u> over a set period of time than an appliance with a <u>lower power rating</u>.

Food Labels **Tell** *You* How Much Energy *is in* **Food**

1) All the <u>food</u> we eat contains <u>energy</u> — it's important to make sure you're taking in the <u>right amount</u> of energy each day (page 7).

2) The energy in food is measured in <u>kilojoules (kJ)</u>.

3) You can <u>compare</u> the amount of <u>energy</u> in different foods by looking at their <u>label</u>.

> You may also see a value for kcals on a food label — this is just another unit that energy can be measured in.

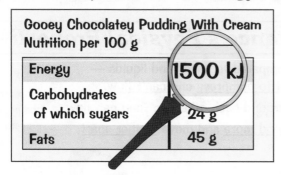

Gooey Chocolatey Pudding With Cream Nutrition per 100 g	
Energy	**1500 kJ**
Carbohydrates of which sugars	24 g
Fats	45 g

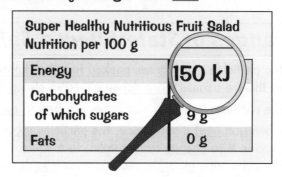

Super Healthy Nutritious Fruit Salad Nutrition per 100 g	
Energy	**150 kJ**
Carbohydrates of which sugars	9 g
Fats	0 g

What's my power rating you ask? About 8 out of 10...

Reading <u>labels</u> is not just good fun, but also useful — a label can tell you the power rating of an <u>appliance</u>, or how much <u>energy</u> is in your <u>food</u>. And, you know what they say — energy is what makes the world go round. Ok, I might have made that up, but this stuff's <u>important</u> — make sure you <u>learn</u> it good 'n' proper.

Physical Changes

Even just a <u>change of temperature</u> can change a substance <u>physically</u>. Sounds easier than the gym...

Physical Changes *Don't Involve a Change in Mass*

1) A substance can either be a <u>solid</u>, a <u>liquid</u> or a <u>gas</u>. These are called <u>states of matter</u>. When a substance <u>changes</u> between these physical states, its <u>mass doesn't change</u>.

2) <u>Physical changes</u> are <u>different</u> to chemical changes because there's <u>no actual reaction</u> taking place and <u>no new substances</u> are <u>made</u>. The <u>particles</u> stay the <u>same</u>, they just have a <u>different arrangement</u> and amount of <u>energy</u>.

There are several <u>different processes</u> that can change the physical state of a substance:

Melting, Evaporating, Condensing, Freezing

1) If you <u>melt</u> a certain amount of <u>ice</u>, you get the <u>same amount</u> of <u>water</u> — and then if you <u>boil</u> that so it <u>evaporates</u>, you get the <u>same amount</u> of <u>steam</u>.

2) It's the same in the <u>other direction</u> — if the <u>steam condenses</u>, you get the <u>same amount</u> of <u>water</u>, and if the <u>water freezes</u>, you get the <u>same amount</u> of <u>ice</u>.

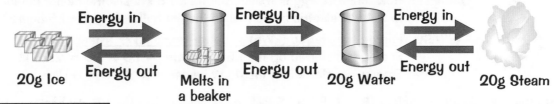

Sublimation

Some substances, such as carbon dioxide, can go <u>straight</u> from being a <u>solid</u> to being a <u>gas</u> — this is called <u>sublimation</u>. When this happens, the <u>mass</u> of gas is (you guessed it) <u>the same</u> as the <u>mass</u> of the solid.

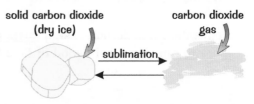

Dissolving

1) When a solid substance <u>dissolves</u> to form a <u>solution</u> (page 39), there's <u>no change</u> in <u>mass</u>. The amount of substance after dissolving is the <u>same</u> as before, it's just in a <u>different form</u>.

2) <u>Dissolving</u> is <u>reversible</u> — if you <u>evaporate</u> all the solvent, you'll be left with the <u>same amount</u> of <u>solid</u> as before it dissolved.

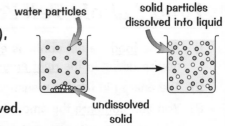

Changes of State *Affect a Substance's Physical Properties*

1) The particles in <u>solids</u> are <u>packed together tightly</u> compared to gases and liquids — so they're usually <u>more dense</u>. They're also <u>difficult to compress</u> and <u>can't flow</u>.

2) The particles in liquids and gases are <u>free to move</u> around each other, so they can <u>flow</u>.

3) When you <u>heat</u> a substance, the particles <u>move around more</u> and move <u>further apart</u>, causing it to <u>change</u> from a solid, to a liquid, to a gas. The substance <u>expands</u> and becomes <u>less dense</u>.

4) <u>Ice</u> is a funny one though — when it <u>melts</u> (to become water), the particles actually come <u>closer together</u> and its <u>density increases</u>.

See p.32 for more on the physical properties of solids, liquids and gases.

All this studying's getting me in a right change of state...

Remember, physical changes have <u>no effect</u> on the <u>chemical structure</u> of the substance — they just cause it to <u>change state</u> between solid, liquid and gas. Ice is really just water, but with the particles in a more structured formation. If it melts, it goes back to being tasty refreshing water without losing any mass at all.

Movement of Particles

Brownian motion has nothing to do with colours — Brown was just the name of the guy who first noticed it.

Brownian Motion *is the Random Movement* of Particles

1) In 1827, a scientist called Robert Brown noticed that tiny pollen particles moved with a zigzag, random motion in water.

2) This type of movement of any particle suspended ('floating') within a liquid or gas is known as Brownian motion.

3) Large, heavy particles (e.g. smoke) can be moved with Brownian motion by smaller, lighter particles (e.g. air) travelling at high speeds — which is why smoke particles in air appear to move around randomly when you observe them in the lab.

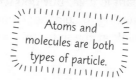
Atoms and molecules are both types of particle.

Diffusion *is Caused by the* Random Motion *of Particles*

1) The particles in a liquid or gas move around at random.

2) Particles eventually bump and jiggle their way from an area of high concentration to an area of low concentration. They constantly bump into each other, until they're evenly spread out across the substance.

High concentration of spectacle-particles   Low concentration of spectacle-particles

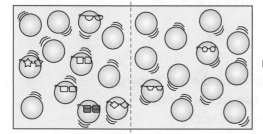

The particles move about randomly until...

...there's an even concentration of spectacle-particles on each side.

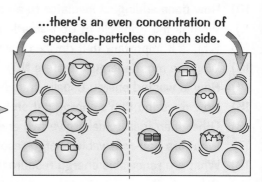

Movement *of Particles* Increases With *Temperature*

1) An increase in temperature causes particles to move around more — their speed increases. This means the spaces between the particles get bigger so they take up more space, which causes the material to expand.

2) If you heat a gas or liquid inside a container, the pressure inside the container increases as the particles bump into the sides more often and more quickly.

3) The particles don't get bigger — they just need more space to move around in.

EXAMPLE:
When the liquid in this flask is heated its volume expands as the particles move apart with their extra energy.

So the liquid moves up the thin tube — this is how a thermometer works — dead useful...

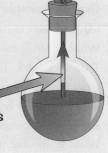

I'm a golf comedian — I specialise in par-tickle movement...

...with a PhD in Greenian motion. Ok, I'll admit that was a fairway off being funny, but it's worth taking a long shot every now and then in case you hit the hole in one. You just have to putt in the time if you want to remember all this stuff, so get in your study bunker and buckle down... it'll be over soon, I promise.

Section Summary

Ah, the section summary — on the home stretch at last. In this section, you got to go on a voyage of discovery into the weird and wonderful world of all things energy and matter. All that's left for you now is to work through the exciting questions below and claim your free ice cream at the start of the next section. Ok, the ice cream is a lie, but you won't regret taking the time to work through these questions if you want to be super-amazing at science, trust me. Take your time with them if you like, and maybe have the odd cheeky peek back at the appropriate page if you're stuck — I won't tell anyone, honest.

1) Name the seven types of energy stores and the four ways of transferring energy between them.
2) Give one example of energy being transferred electrically to one or more energy stores.
3) Describe the energy transfers for an object that is falling.
4) When does an object have energy stored in its chemical energy store?
5) Replace each row of bumble bees in the sentence below with the correct words.
   When a 🐝-🐝-🐝-🐝 moves an object through a distance, 🐝-🐝-🐝-🐝 is transferred.
6)* A crane applies a force of 2000 N to lift a small elephant 10 m. How much energy does it transfer?
7) What does thermal equilibrium mean?
8) Name two ways in which energy can be transferred between two objects by heating.
9) Describe how energy is transferred when two objects at different temperatures are touching.
10) How does adding an insulator to an object affect the rate of energy transfer?
11) What is the Principle of Conservation of Energy?
12) Why is it important that devices transfer energy from one store to another?
13) Why are most energy transfers **NOT** perfect?
14) How is wasted energy usually transferred?
15) How does the Sun's energy get stored in fossil fuels?
16) Other than fossils fuels, give two energy resources created using the Sun's energy.
17) What are non-renewable energy resources? How can we use less of them?
18) What are renewable energy resources? Why will they never run out?
19)* Calculate the energy transferred by a 1.5 kW remote-control gibbon used for half an hour.
20) What unit is household electricity measured in?
21)*Electricity costs 15p per kWh. Calculate the cost of an electricity bill for 298.2 kWh.
22) What does the power rating of an appliance tell you?
23)*Which will transfer more energy — a 200 W device left on for 1 hour, or a 300 W device left on for 1 hour?
24) What unit is the energy in food usually measured in?
25) How could you find out the energy contained in a packet of chocolate-flavoured baba ganoush?
26) Give an example of a physical change of state.
27)*50 g of iron is melted. How much liquid iron would be produced?
28) What is sublimation?
29) True or false? Dissolving is not a reversible process.
30) Give two differences between the physical properties of a gas and a solid of the same substance.
31) What's meant by Brownian motion?
32) Particles in gases move from areas of high concentration to low concentration.
    What is the name of this process?
33) Explain why gases expand when they're heated.

*Answers on page 108.

Speed

Neeeeoww... Yes, it's a page on speed. Make sure you can do these <u>calculations</u> — don't <u>zoom through</u>.

Speed is How Fast You're Going

A formula triangle

1) <u>Speed</u> is a <u>measure</u> of how <u>far</u> you travel in a <u>set</u> amount of <u>time</u>.

2) The <u>formula triangle</u> is definitely the <u>best</u> way to do <u>speed calculations</u>.

$$Speed = \frac{Distance}{Time}$$

This line means divided by or shared by (÷).

3) Use the word <u>SIDOT</u> to help you remember the formula:

<u>SIDOT</u> — <u>S</u>peed <u>I</u>s <u>D</u>istance <u>O</u>ver <u>T</u>ime.

<u>Always</u> use <u>UNITS</u>.

4) There are <u>three</u> common <u>units</u> for speed. You should realise that they're all kind of the same, i.e. <u>distance unit</u> per <u>time unit</u>.

<u>metres</u> per <u>second</u> — m/s
<u>miles</u> per <u>hour</u> — mph or miles/h
<u>kilometres</u> per <u>hour</u> — km/h

Work Out Speed Using Distance and Time

To work out <u>SPEED</u> you need to know the <u>distance travelled</u> and the <u>time taken</u>.

Example 1:

A hooligan sheep is skateboarding down a farmer's track. You notice it takes exactly <u>5 seconds</u> to move between two fence posts, <u>10 metres</u> apart. <u>What's the sheep's SPEED?</u>

10m

Answer:

<u>STEP 1)</u>   <u>Write down what you know:</u>
distance, d = 10 m   time, t = 5 s

<u>STEP 2)</u>   <u>We want to find speed</u>, s
from the formula triangle: s=d/t
Speed = Distance ÷ Time   = 10 ÷ 5   = <u>2 m/s</u>

Put your finger over "S" in the formula triangle — which leaves d/t (i.e. d÷t).

Speed questions are a <u>doddle</u> if you just learn the <u>formula triangle</u>.

Example 2:

A campervan sprints down the motorway and travels 15 miles in 30 minutes. <u>What's its speed?</u>

Answer:

<u>STEP 1)</u>   <u>Write down what you know:</u>
distance, d = 15 miles   time, t = 30 minutes = 0.5 of an hour.

<u>STEP 2)</u>   <u>We want to find speed</u>, s, from the formula triangle: s=d/t
Speed = Distance ÷ Time   = 15 ÷ 0.5   = <u>30 miles/hour (mph)</u>

For the <u>answer</u> to be in <u>miles per hour</u> you need the <u>distance</u> in <u>miles</u> and the <u>time</u> in <u>hours</u> so the 30 mins had to become 0.5 hrs.

Speed is ace — well it takes some beating...

Speed is a pretty simple idea really. This page has all the <u>really basic</u> and important facts about <u>speed</u>. There's the formula for a start, and the units, and then a couple of worked examples. <u>Learn it all</u>. Now.

More on Speed

You know what they say — more haste, less speed. Here's some <u>more speed</u>. Make sure you're not hasty.

Distance-Time **Graphs**

A distance-time graph shows the <u>distance</u> travelled by an object over <u>time</u>.

1) The <u>slope</u> of the line (<u>gradient</u>) shows the <u>speed</u> at which the object is moving.

2) The <u>steeper</u> the graph, the <u>faster</u> the object is going.

3) <u>Flat</u> sections are where it's <u>stopped</u>.

4) <u>Downhill</u> sections mean it's <u>moving back</u> toward its starting point.

5) <u>Curves</u> represent a <u>changing</u> speed.

6) A <u>steepening</u> curve means the object is <u>speeding up</u> (<u>accelerating</u>).

7) A curve <u>levelling off</u> means the object is <u>slowing down</u> (<u>decelerating</u>).

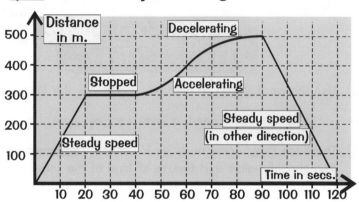

Relative Motion — *Two Objects Passing Each Other*

<u>Relative motion</u> is useful if you want to know the <u>speed</u> of something when <u>you are moving too</u>.

1) If two objects are moving <u>in opposite directions</u> on the <u>same straight line</u> you can <u>add their speeds together</u> to calculate their <u>relative motion</u>. Look:

30 km/h → Relative speed 30 km/h + 25 km/h = 55 km/h ← **25 km/h**

- Both trains are moving <u>towards</u> each other <u>from opposite directions</u>. So if you're sat on the red train, the blue train is getting <u>closer much faster</u> than if you were sat still at the side of the track. This is because it is <u>moving towards you</u> while <u>you're moving towards it</u>.

- To work out the speed of the blue train relative to the red train, just <u>add the speeds together</u>. 30 + 25 = <u>55 km/h</u>, so the speed of the blue train relative to the red train is 55 km/h.

2) If the objects are moving in the <u>same direction</u> on the same straight line you can <u>subtract their speeds</u> to calculate their <u>relative motion</u>.

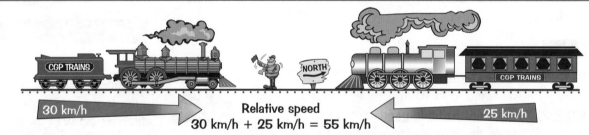

- The car is moving in the <u>same direction</u> as the tractor but at a <u>faster speed</u>. If you're in the car, you're getting further away from the tractor <u>more slowly</u> than if it wasn't moving (since it's <u>moving towards you</u> while <u>you're moving away from it</u>).

  20 mph → Relative speed 30 mph – 20 mph = 10 mph → **30 mph**

- To work out the speed of the car relative to the tractor, <u>subtract the speeds</u>. 30 – 20 = <u>10 mph</u> — the car gets 10 miles further away from the tractor every hour.

I hate Christmas — all my relatives get too motional...

There are <u>two really important things</u> on this page — distance-time graphs and relative motion. Make sure you can read <u>all</u> the info hidden in a <u>distance-time graph</u>, then check you've got <u>relative motion</u> nailed.

Forces and Movement

Well, I can't <u>force</u> you to read this page — but if I were you, I'd <u>push</u> on with it...

Forces are Nearly Always Pushes and Pulls

1) Forces are <u>pushes</u> or <u>pulls</u> that occur when two objects <u>interact</u>.
2) Forces <u>can't</u> be seen, but the <u>effects</u> of a force <u>can</u> be seen.
3) Forces are measured in <u>newtons</u> — <u>N</u>.
4) They usually act in <u>pairs</u>.
5) They <u>always</u> act in a <u>certain direction</u>.
6) A <u>newton meter</u> is used to <u>measure</u> forces.

Objects <u>don't</u> need to <u>touch</u> to interact. The <u>gravitational pull</u> <u>between planets</u> (p.102), <u>forces between magnets</u> (p.99) and forces due to <u>static electricity</u> (p.98) are all <u>non-contact forces</u>.

Forces Can Make Objects Do Five Things

1. <u>Speed</u> Up or <u>Start</u> Moving	Like <u>kicking</u> a football. To <u>start</u> something moving, a push force must be <u>larger</u> than <u>resisting forces</u> like friction (see next page).	3. Change Direction	Like hitting a <u>ball</u> with a <u>bat</u> or gravity causing footballs to come back down to Earth.
		4. Turn	Like <u>turning</u> <u>a spanner</u>.
2. <u>Slow</u> Down or <u>Stop</u> Moving	Like <u>drag</u> or <u>air resistance</u> (see next page).	5. Change Shape	Like <u>stretching</u> and <u>compressing</u> (see p.83), <u>bending</u> and <u>twisting</u>.

Learn These Two Important Statements:

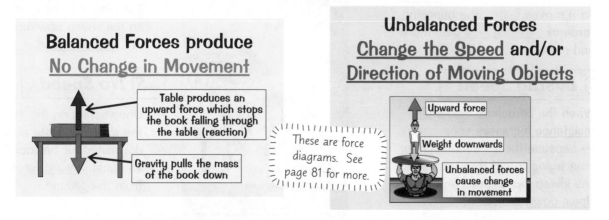

Balanced Forces produce <u>No Change in Movement</u>

Table produces an upward force which stops the book falling through the table (reaction)

Gravity pulls the mass of the book down

These are force diagrams. See page 81 for more.

Unbalanced Forces <u>Change the Speed</u> and/or <u>Direction of Moving Objects</u>

Upward force

Weight downwards

Unbalanced forces cause change in movement

Force is ace — well it beats speed...

Forces are a <u>simple enough</u> idea, but you still need to know <u>all of the details</u> on this page.
Luckily there really isn't much to learn here — just make sure you do. <u>Learn</u>, <u>cover</u>, <u>scribble</u>, <u>check</u>...

Friction and Resistance

Friction Tries to Stop Objects Sliding Past Each Other

Friction is a <u>force</u> that always acts in the <u>opposite</u> direction to movement.
It's the force you need to <u>overcome</u> when <u>pushing an object</u> out of the way.

The Good Points of Friction — It Allows Things to Start and Stop

1) Friction allows the tyres on a bike to <u>grip</u> the road <u>surface</u> — without this grip you couldn't make the bike move <u>forward</u> and you wouldn't be able to <u>stop</u> it either — it'd be like riding on <u>ice</u>.

2) Friction also acts at the <u>brakes</u> where they <u>rub</u> on the <u>rim</u> of the <u>wheel</u> or on the <u>brake disc</u>. Friction also lets you <u>grip</u> the <u>bike</u> — important if you want to ride it without slipping off.

The Bad Points of Friction — It Slows You Down

1) <u>Friction</u> always <u>wastes energy</u> — friction between the moving parts of a bike <u>warms up</u> the gears and bearings — a <u>waste</u> of energy.

2) Friction <u>limits top speed</u>. The <u>air resistance</u> (a kind of friction, see below) takes <u>most</u> of your energy and <u>limits</u> your maximum <u>speed</u>.

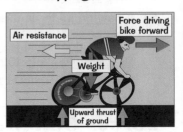

Air and Water Resistance Slow Down Moving Objects

1) Air and water resistance (or "drag") <u>push against</u> objects which are moving through the air or water.

2) These are kinds of <u>frictional</u> force because they try to <u>slow</u> objects down.

3) If things need to go fast, then they have to be made very <u>streamlined</u> — which just means they can <u>slip</u> through the <u>air</u> or <u>water</u> without too much resistance. A good example is a sports car.

How Air Resistance Affects Sheep Jumping Out of Planes

(It happens all the time round here.)

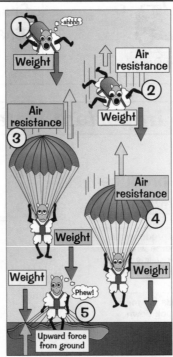

1) Gains Speed

At the start, the sheep only has the <u>force</u> of its <u>weight</u> (i.e. <u>gravity</u>) pulling it down — so it starts to <u>move faster</u>.

2) Still Gaining Speed

As it moves <u>faster</u>, the opposing force of <u>air resistance</u> gets more and more.

3) Losing Speed

When the parachute opens <u>air resistance increases</u> enormously — because there's a much <u>larger</u> area trying to <u>cut</u> through the air. The sheep loses speed and <u>slows down</u> gratefully.

4) Steady Speed

Very quickly the <u>air resistance</u> becomes <u>equal</u> to the <u>weight</u> — the two forces are <u>balanced</u>. The overall force is zero, so the sheep now moves at a <u>steady speed</u>.

5) No Speed

Once safely on the ground, the sheep's <u>weight</u> acting downwards is balanced by an equal <u>upward force</u> from the ground.

Resistance is useless — you've just got to learn about friction...

Air resistance — what a drag, huh. Anyway, here's a page of <u>key facts</u> that you need to <u>learn</u>. Go go go.

Force Diagrams

Force diagrams. They're <u>diagrams</u> that show <u>forces</u>. Bet you weren't expecting that...

Show the Forces Acting on an Object Using a Force Diagram

Force diagrams show the <u>forces</u> acting on an object and whether they are <u>balanced</u> or <u>unbalanced</u>.

Example: Stationary Teapot Force Diagram

Here's a teapot on a table...

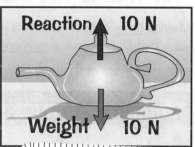

1) The force of <u>gravity</u> (or weight) is acting <u>downwards</u> on the teapot (see page 102) — it's the red arrow.

2) This causes a <u>reaction force</u> from the table's surface <u>pushing up</u> on the teapot — the <u>blue</u> arrow.

3) The <u>reaction force</u> and <u>weight</u> are <u>equal</u> and <u>opposite</u> — you can tell because the <u>arrows are the same size</u> and pointing in <u>opposite directions</u>.

4) This means the <u>forces</u> on the teapot are <u>BALANCED</u>. So it <u>remains stationary</u> (not moving).

If the forces acting on a <u>moving object</u> are <u>balanced</u>, it <u>carries on moving</u> at a <u>steady speed</u> in the <u>same direction</u>.

Example: Accelerating Bus Force Diagram

Here's a force diagram of a bus...

1) The <u>red</u> arrow shows that the engine is creating a force of 2000 N to make the bus move <u>forwards</u>.

2) The <u>blue</u> arrow shows that there is a <u>frictional</u> force of 500 N acting in the <u>opposite direction</u>.

3) The forces are <u>unbalanced</u> (the arrows in the diagram are <u>unequal</u> sizes) so the bus is <u>accelerating</u> in the direction of the <u>bigger force</u> (forwards).

You Can Add or Subtract Forces Along the Same Line

1) If you've got a force diagram where all the forces are acting along the same line (e.g. <u>forwards and backwards</u> OR <u>up and down</u>), you can calculate the <u>overall force</u> by <u>adding</u> or <u>subtracting</u> the forces.

2) This is handy for <u>working out</u> if an object is <u>accelerating</u> (getting faster), <u>decelerating</u> (slowing down) or staying at a <u>steady speed</u>:

Forces acting along the same line are said to be acting in <u>one dimension</u>.

Golden Rules of Force Diagrams

1) If the forces are acting in <u>opposite directions</u>, you <u>subtract</u> the forces to get the <u>overall force</u>.

2) If they're acting in the <u>same direction</u>, you <u>add</u> the forces together to get the <u>overall force</u>.

Overall force	Overall force	Overall force
100 – 100 = 0 N	100 – 500 = –400 N	200 + 50 – 20 = 230 N
No acceleration, moves at a steady speed	Strong deceleration	Strong acceleration

Use the force — look at a diagram, find out if it's accelerating...

Just remember: if the arrows on a force diagram are <u>equal</u> and <u>opposite</u>, the object <u>isn't moving</u> or is moving at a <u>steady speed</u>. If they're <u>different sizes</u>, the forces are <u>unbalanced</u> and the <u>speed</u> will <u>change</u>.

Moments

Don't wait a lifetime to learn moments like this — <u>memorise</u> what's on this page <u>now</u>.

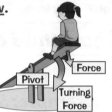

Forces **Cause** *Objects* **to** *Turn* **Around** *Pivots*

A <u>pivot</u> is the point around which rotation happens — like the middle of a seesaw.

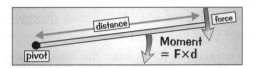

A **Moment** *is the* **Turning Effect** *of a Force*

1) When a <u>force acts</u> on something which has a <u>pivot</u>, it creates a <u>moment</u>.

2) Learn this important equation:

Moment = force x perpendicular distance

in newton metres, Nm | in newtons, N | in metres, m

$$M = F \times d$$

> You might remember moments from 'How Muscles Work' (p. 11). Sadly, you need to revisit them again here.

Balancing **Moments**

Balanced moments mean that...

anticlockwise moments = clockwise moments

> If the moments are <u>not</u> balanced, the ruler will turn in the direction of the bigger moment.

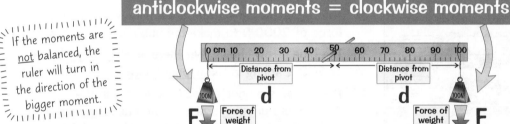

Clockwise

<u>ANTICLOCKWISE</u>   force x perpendicular distance = force x perpendicular distance   <u>CLOCKWISE</u>

100 N x 0.5 m = 100 N x 0.5 m

<u>50 Nm</u> = <u>50 Nm</u>

— BALANCED

Is it **Balanced?**

Which rulers are balanced? If you think the ruler is <u>balanced</u> write it on a bit of paper. If you reckon it's unbalanced, then write <u>unbalanced</u>, but say <u>which side</u> of the ruler will dip down. Words to use: balanced, unbalanced, left side down or right side down. Answers on page 108.

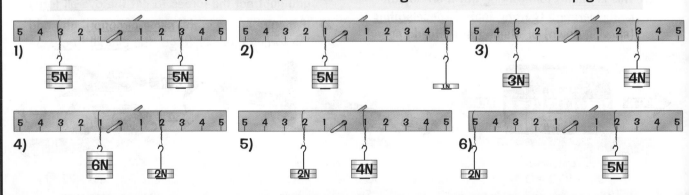

Learn all this page — it'll only take a few moments...

Nope — I don't know why they call them "<u>moments</u>" either. It's a good word for a short period of time, but that's about it. After that it just seems to cause <u>confusion</u>. But there you go — it wouldn't be <u>proper</u> science if it all made perfect sense first time round. And it's a better name than <u>rotating forcey thingy</u>.

Forces and Elasticity

It's not just about turning, pushing and pulling — forces are also able to <u>stretch</u> or <u>squash</u> things.

You Can Deform Objects by Stretching or Squashing

1) You can use forces to <u>stretch</u> or <u>compress</u> (squash) objects, e.g. when you stomp on an empty fizzy pop can.
2) The force you apply causes the object to <u>deform</u> (change its shape).
3) <u>Springs</u> are <u>special</u> because they usually <u>spring back</u> into their <u>original shape</u> <u>after</u> the force has been <u>removed</u> — they are <u>elastic</u>.

Work is Done When a Force Deforms an Object

1) You might remember <u>energy transfer</u> from <u>page 67</u> (if not, take a look). <u>Work done</u> is the <u>same thing</u>.
2) Energy is transferred and work is done when an object is <u>deformed</u>. For example:

- When you <u>stretch</u> a <u>spring</u>, you're <u>doing work</u> by transferring <u>energy</u>.
- The energy is transferred from the <u>kinetic energy store</u> of the spring to its <u>elastic energy store</u>.
- When the spring 'springs' back into its <u>original shape</u>, the energy is <u>transferred</u> back to the <u>kinetic energy store</u>.

There's more on different stores of energy on page 66.

Hooke's Law Says Extension of a Spring Depends on the Force

If a spring is supported at the top and then a weight is attached to the bottom, it <u>stretches</u>.

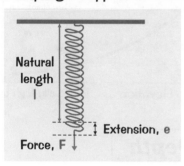

Natural length
l

Extension, e
Force, F ↓

1) <u>Hooke's Law</u> says the amount it stretches (the <u>extension</u>, <u>e</u>), is <u>directly proportional</u> to the <u>force applied</u>, F. I.e. the relationship between force and extension is <u>linear</u>.
2) Some objects <u>obey</u> Hooke's Law, e.g. <u>springs</u>. But it <u>only</u> applies up to a <u>certain force</u>.
3) For springs, the force at which Hooke's Law <u>stops working</u> is <u>much higher</u> than for most materials. Springs are <u>unusual</u>.

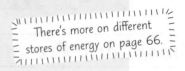

Hooke's Law
$$F = k \times e$$

k is the spring constant. Its value depends on the material that you're stretching and it's measured in newtons per metre (N/m).

When a Stretched Spring Holds a Weight, it's in Equilibrium

1) <u>Equilibrium</u> is just a fancy way of saying <u>all the forces are balanced</u>.
2) When a <u>stretched</u> or <u>compressed</u> spring holds a weight <u>still</u>, the force of the weight is <u>the same</u> as the force of the spring as it tries to return to its original shape. So the forces are balanced and in <u>equilibrium</u>.

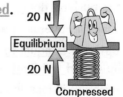
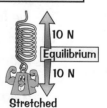

20 N — Equilibrium — 20 N — **Compressed**
10 N — Equilibrium — 10 N — **Stretched**

This page will stretch you — better do some extra work on it...

It's true — there's an awful lot of stuff to learn here. So read it through <u>again</u>, cover up the page and see if you can <u>scribble</u> down the headings. Then the <u>diagrams</u> and <u>equation</u>. Then the <u>rest</u>. Better get to it...

Pressure

Don't let pressure get you down — here's a lovely page that explains it all. Now that's a load off your mind.

Pressure is How Much *Force* is Put on a Certain *Area*

Pressure, force and area are all kind of tied up with each other — as the formula shows. The formula can also be put in a triangle, which is nice.

A given force acting over a big area means a small pressure (and vice versa).

$$\text{Pressure} = \frac{\text{Force}}{\text{Area}}$$

Pressure is Measured in N/m^2 or Pascals *(Pa)*

If a force of 1 newton is spread over an area of 1 m^2 (like this) then it exerts a pressure of 1 pascal. Simple as that.

1 newton/metre^2 = 1 pascal
1 N/m^2 = 1 Pa

Force acts normal (at 90°) to area.

Pressure =1Pa

Atmospheric Pressure is *All Around Us* All the Time

The weight of the atmosphere is constantly pushing against us — but we're so used to it we can't feel it.

1) The lower you are, the more atmosphere there is above you — so the pressure due to the weight of the atmosphere increases.

2) If you gain height, there's less atmosphere above you, so the atmospheric pressure decreases.

3) Atmospheric pressure is over 100 000 Pa at sea level. But at the top of Mt Everest (8800 m above sea level) the atmospheric pressure is only around 33 000 Pa.

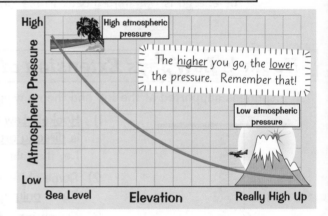

High atmospheric pressure

The higher you go, the lower the pressure. Remember that!

Low atmospheric pressure

Atmospheric Pressure — High / Low

Sea Level — Elevation — Really High Up

The Pressure in Liquids Increases *with* Depth

For liquids like water, the pressure increases with depth due to the weight of water above.

Water Pressure Causes Upthrust *and Makes Things* Float

1) If you place an object in water, it experiences water pressure from all directions.

2) Because water pressure increases with depth, the force pushing upwards at the bottom of the object is greater than the force pushing down at the top of the object.

3) This causes an overall upwards force, called upthrust.

4) If the upthrust is equal to the object's weight, then the object will float — like this boat:

5) If the upthrust is less than the object's weight, it will sink.

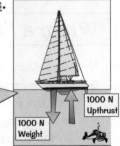

1000 N Upthrust

1000 N Weight

Pressure — pushing down on you, pressing down on me...

First things first, get that formula learnt. Remember: air and water have a mass. If you increase the amount of this mass above your head, you're increasing the force acting over a given area and you'll get an increase in pressure. That's why pressure in liquids increases with depth and why air pressure decreases with height.

Section Summary

Section 9 is all about forces and motion. It's all pretty straightforward stuff really and the questions below will test whether you've learnt the basic facts.

If you're having trouble learning the stuff, try taking just one page on its own. Start by learning part of it, then covering it up and scribbling it down again. Then learn a bit more and scribble that down. Soon enough you'll have learnt the whole section and be ready to face whatever questions your teachers throw at you.

1) What exactly is speed? Write down the formula triangle for speed.

2) How does SIDOT help you remember what speed is?

3)* A bogie is flicked across the lab by some hoodlum. It travels 5 m in 2 seconds. Calculate the speed of the bogie.

4)* On sports day you run 100 m in 20 seconds. Can you run faster than the flicked bogie?

5)* When a car is going at 40 mph, how far will it travel in 15 minutes?

6) What does the gradient show on a distance-time graph?

7) What does a straight, flat line mean on a distance-time graph?

8) How would you calculate the relative speed of two trains travelling in the same direction?

9) Can forces be seen? How do we know they're there?

10) What are the units of force? What would you use to measure force?

11) What are the five different things that forces can make objects do?

12) What do balanced forces produce? What do unbalanced forces do?

13) What is friction? When does it occur?

14) Give three good points of friction. Give two bad points of friction.

15) What is air resistance? And water resistance?

16) When a sheep first jumps out of a plane what happens to its speed?

17) As the sheep moves faster, what happens to the air resistance?

18) What happens to air resistance when the sheep's parachute opens?

19) Does the speed then change? When does the sheep's speed become steady?

20) What might happen if the ground didn't provide an upward force to equal the sheep's weight?

21) Draw a force diagram of a kettle resting on a table. The force due to gravity acting on the kettle is 10 N.

22) If the forces acting on a moving bus are balanced, what will happen to its speed?

23) Draw a force diagram showing a bus accelerating. Make sure the arrows are different sizes.

24) What is a pivot?

25) What is a moment? Give the formula for a moment.

26) What does "balanced moments" mean?

27) *A force of 100 N is put 1 m away from the middle of a seesaw. What distance from the middle should a force of 50 N be applied to balance the seesaw?

28) Give two ways you can deform objects.

29) What does Hooke's Law say? Write down the formula.

30) What is pressure? Give the formula for calculating pressure.

31)* A force of 200 N acts on an area of 2 m². Calculate the pressure.

32) Is atmospheric pressure higher at the seaside or up a mountain? Why?

33) When does an object placed in water float?

*Answers on page 108

Water Waves

Take a deep breath and dive into the wonderful world of water waves...

Water Waves are Transverse

1) Waves travelling across the ocean are good examples of transverse waves.

2) A transverse wave has undulations (up and down movements) that are at right angles to the direction the wave is travelling in.

3) Waves transfer energy from one place to another. So the undulations are also at right angles to the direction of energy transfer.

4) Lots of other important waves are transverse too, like light (see next page).

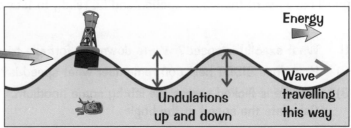
Undulations up and down | Wave travelling this way | Energy

Waves Can be Reflected

1) If a water wave hits a surface, it will be reflected.

2) This causes the direction of the wave to change.

3) All waves can be reflected.
There's more on reflection on page 88.

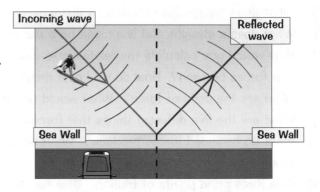
Incoming wave | Reflected wave | Sea Wall | Sea Wall

Transverse Waves Have Crests, Troughs and Displacement

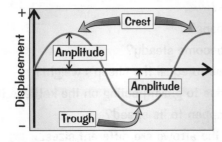
Displacement | Crest | Amplitude | Amplitude | Trough

1) The crest is the highest part of the wave.

2) The trough is the lowest part of the wave.

3) The displacement is how far a point on the wave is from the middle line. The amplitude is the maximum displacement — the distance from the middle of the wave to a crest or trough.

Superposition Happens When Two Waves Meet

1) If two water waves meet, their displacements combine briefly. This is superposition. The waves then carry on as they were.

2) If two identical CRESTS meet, the height of each wave is added together. So the crest height doubles.

3) If two identical TROUGHS meet, the depth of each wave is added together. So the trough depth doubles.

4) If one wave is at a crest and the other is at a trough, you subtract the trough depth from the crest height. So the crest or trough will be smaller and may even cancel out, leaving a flat water surface.

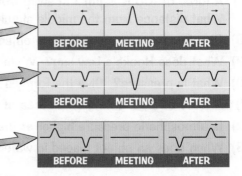
BEFORE | MEETING | AFTER

Don't let your progress waver — reflect on what you've learnt...

Eeek, that was a bit complicated. So grab a piece of paper and write down all the technical terms on this page and what they mean. Then draw and explain the diagrams. Finally, do all that again with the book closed.

Light Waves

You wouldn't know from looking, but light is actually a wave. Here's a page all about it...

Light is a Wave that Transfers Energy

1) Light is produced by luminous objects such as the Sun, candles, light bulbs, flames and glow worms.
2) Light is a wave, which always travels in a straight line.

Light Waves and Water Waves Are Similar...

1) Like waves in water, light waves are transverse waves — they have undulations at right angles to the direction the wave is travelling in (see previous page).

2) And like waves in water, light waves transfer energy from one place to another.

3) Light waves can be reflected too — this is how mirrors work (see next page for more).

...But Light Waves Don't Need Particles to Travel

1) Water waves travel (and transfer energy) by moving particles.
2) Light waves don't need particles to travel. This is a good thing — light from the Sun has to travel through space (where there aren't many particles, see below) to reach Earth.
3) Light waves are slowed down by particles.

Light Waves Always Travel at the Same Speed in a Vacuum

1) Light travels faster when there are fewer particles to get in the way.

2) Light always travels fastest in a vacuum. A vacuum is where there is nothing at all — no air, no particles, nothing. Space is mostly a vacuum.

3) The speed of light in a vacuum is always 3×10^8 m/s (that's three hundred million metres per second). It's a constant.

4) This means light from the Sun gets to Earth in only 8.3 minutes — even though it's 150,000,000 km away.

5) Nothing travels faster than light in a vacuum.

6) Make sure you learn this:

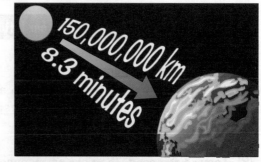

150,000,000 km
8.3 minutes

> Speed of light waves in a vacuum = 3×10^8 m/s

7) Although light is slower when it has to travel through matter (like air or water), it's still so fast that its movement appears instant to the human eye.

Confused? Let me shed some light on the problem...

So there you have it. Light is just like all those waves you see at the beach. Except that it doesn't need a load of water to get from A to B — anything like water puts a load of particles in the way and slows the light waves down. Nope, light only hits top gear when it's in a vacuum with absolutely nothing in the way at all.

Reflection and Refraction

Take a moment and <u>reflect</u> on what you're about to learn...

Mirrors Have Shiny Surfaces Which Reflect Light

1) A light wave is also known as a light <u>ray</u>.
Light rays <u>reflect</u> off <u>mirrors</u> and <u>most other things</u>.

2) <u>Mirrors</u> have a very <u>smooth shiny surface</u>, which reflects all the light off at the <u>same angle</u>, giving a <u>clear reflection</u>. This is <u>specular reflection</u>.

3) <u>Rough surfaces</u> look <u>dull</u>, because the light is reflected back (scattered) in lots of different directions. This is <u>diffuse reflection</u> (or <u>diffuse scattering</u>).

4) Learn the <u>LAW OF REFLECTION</u>:

> ANGLE OF INCIDENCE = ANGLE OF REFLECTION
> ANGLE i = ANGLE r

5) The <u>angle of incidence</u> and the <u>angle of reflection</u> are always measured between the <u>light ray</u> and the <u>normal</u>.

6) The <u>normal</u> is a line at a <u>right angle</u> (90°) to the surface.

Not shiny

This is a ray diagram. Make sure you draw straight lines and get the angles the same when drawing ray diagrams (use a ruler and a protractor).

Refraction is When Light Bends as it Crosses a Boundary

1) Light will travel through <u>transparent</u> (see-through) material, but it <u>won't</u> go through anything <u>opaque</u> (not see-through).

2) Any <u>substance</u> that <u>light</u> (or another wave, e.g. sound) <u>travels through</u> is called a <u>medium</u>.

3) When light travels <u>from one</u> transparent medium <u>to another</u>, it <u>bends</u> or <u>refracts</u>.

<u>LEARN THESE REAL GOOD:</u>

The plural of medium is 'media'.

> When light goes from a <u>LESS</u> dense medium to a <u>MORE</u> <u>dense</u> medium: light bends <u>TOWARDS THE NORMAL</u>.

Example: <u>air</u> to <u>glass</u>.

> When light goes from a <u>MORE dense</u> medium to a <u>LESS</u> <u>dense</u> medium: light bends <u>AWAY FROM THE NORMAL</u>.

Example: <u>glass</u> to <u>air</u>.

Light hitting a glass block is like a car hitting sand

1) <u>Light</u> hits the <u>glass</u> at an <u>angle</u>, <u>slows down</u> and <u>bends</u>.

2) It's a bit like a <u>car</u> hitting <u>sand</u> at an angle. The right wheels get <u>slowed down first</u> and this turns the car to the <u>right</u> — <u>TOWARDS</u> the normal.

3) Leaving the sand, the right wheel <u>speeds up</u> <u>first</u> and this turns the car to the <u>left</u> — <u>AWAY from the normal</u>.

4) If <u>both</u> wheels hit the sand <u>together</u> they <u>slow down together</u>, so the car goes straight through, <u>WITHOUT TURNING</u>.

5) <u>Light</u> does exactly the <u>same</u> when it hits the glass block <u>straight on</u>.

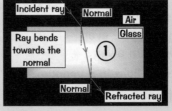

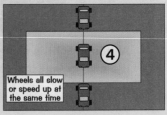

My friend thought she was a medium — but I saw through her...

Reflection and refraction — the first thing you've gotta do is spot that they're actually <u>two different words</u>. Watch, I'll do it again: ref-<u>l-e-c</u>-tion and ref-<u>r-a-c</u>-tion. Now all you need to do is <u>learn</u> what they both are.

How We See

An <u>important</u> page this one — it's all about how we <u>see</u> stuff.

We See Things Because Light Reflects into our Eyes

1) When <u>luminous objects</u> produce light (see page 87), it <u>reflects off</u> <u>non-luminous</u> objects, e.g. you, me, books, sheep, etc.

2) Some of the reflected light then goes <u>into our eyes</u> and that, my friend, is how we see.

The Pinhole Camera is a Simple Camera

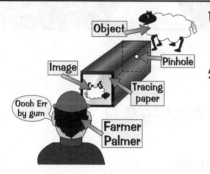

1) The light travels in a <u>straight line</u> from the sheep to the tracing paper <u>screen</u> through the <u>pinhole</u>. Because the hole is small, <u>only one ray</u> gets in from <u>each point</u> on the sheep.

2) The <u>image</u> of the sheep seen by the farmer is <u>upside down</u> and <u>crossed over</u>. This is because the rays of light <u>cross over</u> inside the camera:

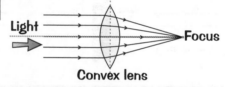

Lenses Can be Used to Focus Light

1) A lens <u>refracts</u> (bends) light.

2) A <u>convex</u> lens <u>bulges outwards</u>. It causes rays of <u>light</u> to <u>converge</u> (move <u>together</u>) to a <u>focus</u>.

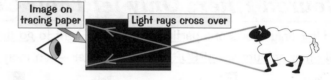

3) In the <u>human eye</u>, the <u>cornea</u> is a transparent 'window' with a <u>convex shape</u>. The cornea does most of the eye's <u>focusing</u>.

4) The convex <u>lens</u> behind the cornea changes shape to focus light from objects at <u>varying distances</u>.

5) The <u>iris</u> is the <u>coloured</u> part of the eye. It <u>controls</u> the <u>amount of light</u> entering the eye.

6) <u>Images</u> are formed on the <u>retina</u>. Cells in the retina are <u>photo-sensitive</u> (sensitive to <u>light</u>).

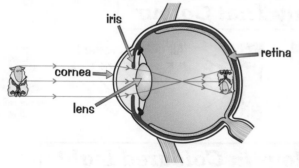

Energy is Transferred From a Light Source to an Absorber

1) <u>Energy</u> is <u>carried</u> by light waves.

2) Anything that <u>absorbs</u> this energy is called an <u>absorber</u>, e.g. a <u>retina cell</u> in the <u>eye</u>, the <u>film</u> in a <u>film camera</u> or the <u>digital image sensor</u> in a <u>digital camera</u>.

3) The energy is <u>transferred</u> to the <u>absorber</u> when it hits the absorber.

4) When light waves hit a <u>retina cell</u> it causes <u>chemical</u> and <u>electrical changes</u> in special cells that send signals to the <u>brain</u>.

5) In a <u>digital camera</u>, light causes the sensor to generate an <u>electrical charge</u>. The changes in charge are read by a <u>computer</u> and turned into an <u>image</u>.

Next up — How We Dee, Eee and Eff...

So that's how we see stuff. Now it needs <u>learning</u> — so cast your gaze back up to the top of the page and read it all over <u>again</u>. Then <u>cover up</u> the book and note down the most important bits.

Colour

Ok, prepare yourself — there's a big plot twist coming up on this page. Hold on to your hats.

White Light is Not just a Single Colour

1) Bit of a shocker, I know — but white light is actually a mixture of colours.

2) This shows up bigstyle when white light hits a prism or a rain drop. It gets dispersed (i.e. split up) into a full rainbow of colours.

3) The proper name for this rainbow effect is a spectrum.

4) Learn the order that the colours come out in:
Red Orange Yellow Green Blue Indigo Violet
Remember it with this historical jollyism:
Richard Of York Gave Battle In Velvet

Prism — A spectrum
White light — Violet is bent the most
red orange yellow green blue indigo violet

5) The frequency of light is the number of complete waves that pass a point per second.

6) Light waves increase in frequency from red (low frequency) to violet (high frequency).

Coloured Filters Only let Their Colour Through

1) A filter only allows one particular colour of light to go through it.

2) All other colours are ABSORBED by the filter — so they don't get through.

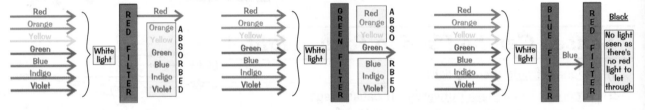

Coloured Objects Reflect Only That Colour

1) Blue jeans are blue because they diffusely reflect blue light and absorb all the other colours.

2) White objects REFLECT all colours.

3) Black objects ABSORB all colours.

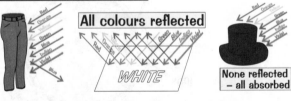

All colours reflected

WHITE

None reflected – all absorbed

Objects Seem to Change Colour in Coloured Light

In white light

White light

1) The boot looks red — it reflects red light and absorbs all other colours.

2) The lace looks green — it reflects green light and absorbs all other colours.

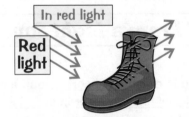

In red light

Red light

1) The boot looks red — it reflects the red light.

2) The lace looks black — it has no green light to reflect and it absorbs all the red light.

In green light

Green light

1) The boot looks black — it has no red light to reflect and it absorbs the green light.

2) The lace looks green — it reflects the green light.

Red boots and green laces? Colour me impressed...

Light — it's full of different colours, but they're all mixed up together so it looks white. Better get learning it.

Sound

Like light, sound is a wave. It's a different type of wave to light though.

Longitudinal Waves Have Vibrations Along the Same Line

1) Longitudinal waves have vibrations that are parallel to the direction of the wave.

2) This means the vibrations are also parallel to the direction of energy transfer.

3) Examples of longitudinal waves include:

   • Sound waves.

   • A slinky spring when you push the end.

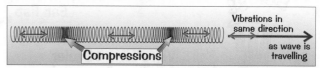

Sound Travels as a Longitudinal Pressure Wave

1) Sound waves are caused by vibrating objects.

2) Sound needs a medium (e.g. air or water) to travel through because something has to pass on the sound vibrations.

3) The vibrations are passed through the medium as a series of compressions (regions of squashed up particles).

4) Sound can't travel in space, because it's mostly a vacuum (there are no particles).

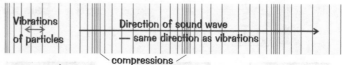

Sound Can be Reflected and Absorbed

1) Sound can be reflected and refracted just like light (see page 88). An echo is sound being reflected from a surface.

2) Sound can also be absorbed. Soft things like carpets, curtains, sheep, etc. absorb sound easily.

The Speed of Sound Depends On What it's Passing Through

1) The more particles there are, the faster a sound wave travels.

2) Dense media have lots of particles in a small space. So the denser the medium, the faster sound travels through it (usually).

3) Sound generally travels faster in solids (like wood) than in liquids (like water) — and faster in liquids than in gases (like air).

4) Sound travels much slower than light.

Frequency is the Pitch of Sound

1) The frequency of sound is the number of complete waves that pass a point per second. A high frequency means more vibrations per second.

2) Frequency is a measure of how high-pitched (squeaky) the sound is. A high frequency means a high-pitched sound.

3) Frequency is measured in hertz (Hz) — the number of vibrations per second.

Dear Coldplay, birds can't fly at the speed of sound. Love, CGP.

Yep. That's another sort of wave for you to get your head around. Sound waves are caused by vibrations — if you've ever put a hand on a bass speaker (or turned up the volume and felt the floor vibrate) you'll have 'felt' a sound wave being made. Higher-pitched sounds are just the same, but the vibrations are more frequent.

Hearing

They say gags about mishearing <u>don't work</u> in books. I think that's silly — I'm not fearing anything. Anyway, let's crack on with this here page...

Sound Waves Make Your Ear Drum Vibrate

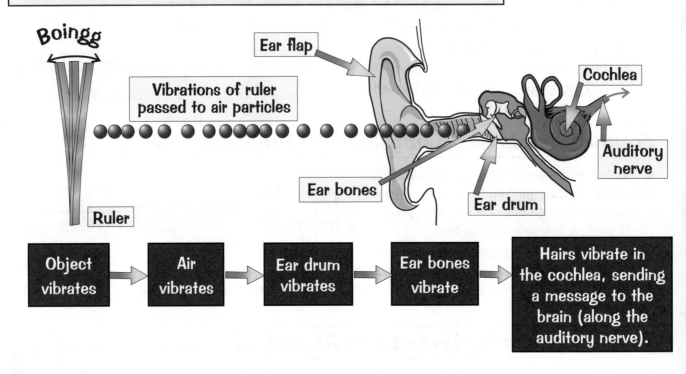

Boingg

Vibrations of ruler passed to air particles

Ruler

Ear flap

Cochlea

Auditory nerve

Ear bones

Ear drum

| Object vibrates | → | Air vibrates | → | Ear drum vibrates | → | Ear bones vibrate | → | Hairs vibrate in the cochlea, sending a message to the brain (along the auditory nerve). |

People and Animals Have Different Auditory Ranges

Your <u>AUDITORY RANGE</u> is the range of <u>frequencies</u> (vibrations per second) that you can <u>hear</u>.

1) The auditory range of humans <u>varies a lot</u> — but it's typically <u>20-20 000 Hz</u>.

2) This means we <u>can't hear low-pitched</u> sounds with frequencies of <u>less</u> than 20 Hz or <u>high-pitched</u> sounds <u>above</u> 20 000 Hz.

3) Some animals like <u>dogs</u>, <u>bats</u> and <u>dolphins</u> can hear much <u>higher frequencies</u> than humans, as the chart shows.

Look back at the previous page for more on frequency and pitch.

Frequency of sound in hertz

120,000
100,000
80,000
60,000
40,000
20,000
0

Save 50% on your auditory range — chop an ear off...

Hey I tell you what, it's a bit of a <u>blockbuster</u> of a <u>diagram</u> that one at the top isn't it? I reckon you really should <u>learn</u> what <u>all</u> the <u>labels</u> are. Remember that we hear things because the air carries the vibrations right into our ear. Also <u>learn</u> all the stuff about <u>auditory ranges</u> — you never know when it might come in useful.

Energy and Waves

You might remember how waves transfer energy (page 86). Well here's a whole page on how useful that is.

Information *Can be Transferred* by Pressure *Waves*

1) All waves transfer energy from one place to another. In doing so, they can also transfer information.
2) Sound waves do this through vibrations between particles — in other words, the pressure changes.
3) This is very useful for recording and replaying sounds.

Sound Waves *are Detected by* Diaphragms *in Microphones*

1) The vibrations in a sound wave make a sensitive diaphragm (e.g. a thin paper or plastic sheet) vibrate inside the microphone.
2) The microphone converts the vibrations to electric signals.
3) Another device can record the electrical signals so that the sound can be reproduced later.

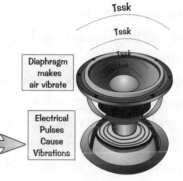

Diaphragm makes air vibrate

Electrical Pulses Cause Vibrations

Loudspeakers Recreate **Sound Waves**

1) An electrical signal is fed into a loudspeaker.
2) This signal causes the diaphragm to vibrate.
3) This makes the air vibrate, producing sound waves.
It's a bit like a microphone in reverse.

Ultrasound *is High Frequency* Sound That We *Can't Hear*

1) Ultrasound includes all sounds that have a higher pitch than the normal auditory range of humans.
2) So that'll be any sound over 20 000 Hz.

Ultrasonic Cleaning *Uses Ultrasound*

High-frequency sound waves are used to clean things — the vibrations of the pressure waves dislodge dirt in tiny cracks that wouldn't normally be cleaned.

1) An item is placed in a special bath filled with water (or another liquid).
2) High-pressure ultrasound waves cause bubbles to form in cavities (holes).
3) The bubbles knock any bits of dirt (contaminants) off the object, leaving it clean enough to eat your dinner off.

You can use ultrasonic cleaning to clean jewellery, false teeth, fountain pen nibs, etc.

Ultrasound *Physiotherapy* May be *Helpful*

1) Ultrasound pressure waves transfer energy through matter — so they can reach inside your body.
2) Some physiotherapists think that this means ultrasound can be used to treat aches and pains in parts of the body that are hard to access — like muscles and tendons deep inside your shoulders.
3) But scientists have found little evidence that ultrasound physiotherapy is an effective treatment.

Ultrasound? What's next, SuperDuperSound?

This page is full of uses for the energy that is transferred by sound waves. And just in case you were wondering, you do have to learn them. Listening to your music player on the way to school doesn't count...

Section Summary

Section 10 tells you everything you need to know about waves. There are quite a few words in there — and some pretty important diagrams too. Science isn't always a complete doddle, so you're bound to find some of the facts a bit tricky to learn. Never fear! As somebody famous once said, "Nothing can take the place of persistence" — in other words, if you want to achieve anything worthwhile or difficult, all you have to do is keep on slogging away at it. Better get cracking with this lot then...

1) What do water waves look like? Sketch out a diagram and label it.
2) Describe what happens to the displacement when two waves meet. What is this called?
3) Give three similarities between water waves and light waves.
4) Give one big difference between water waves and light waves.
5) What speed does light travel at in a vacuum?
6) What is meant by a diffuse reflection?
7) What is the law of reflection?
8) What does that bizarre word "medium" mean?
9) What is refraction?
10) What happens when light goes from a less dense medium to a more dense medium?
11) What happens when light goes from a more dense medium to a less dense medium?
12) Explain in your own words why light "bends" as it enters a glass block.
13) Sketch a diagram of a pinhole camera.
14) Use a diagram to explain why the image is upside down and crossed over.
15) Draw a diagram to show how a convex lens refracts parallel rays of light.
16) Which two parts of the eye help you focus on an object?
17) How do digital cameras form images?
18) How could you show that white light is not just one colour?
19) What is the jollyism for remembering the order of colours in a spectrum?
20) What colour of light has the highest frequency?
21) What colour of light will a red filter let through?
22) Why does something blue look blue in white light?
23) What happens to all the colours in white light when they hit a black object?
24) What colour would green laces look in red light and why?
25) What type of wave are sound (pressure) waves? In which direction are the vibrations?
26) What does sound need to travel from one place to another?
27) Why couldn't you hear a ringing bell in a vacuum?
28) What is an echo?
29) Does sound usually travel faster in solids, liquids or gases? Explain your answer.
30) What does the frequency of a sound mean?
31) Draw a labelled diagram of an ear. Explain how a flicked ruler can be heard.
32) What does auditory range mean?
33) What is the auditory range of humans?
34) How do microphones work?
35) What is ultrasound? What can it be used for?

Electrical Circuits

First up in this section, some <u>electricity basics</u>...

Electric Current *is the Flow of Charge*

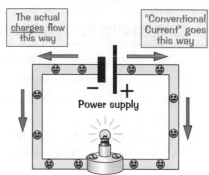

The actual charges flow this way

"Conventional Current" goes this way

Power supply

1) <u>Electric current</u> is the <u>flow</u> of <u>charge</u> around a circuit.

2) It can only flow if a circuit is <u>complete</u>.

3) The moving charges are actually <u>negative electrons</u> (page 98).

4) Irritatingly, they flow the <u>opposite</u> way to the direction of <u>conventional current</u>, which is shown on circuits as <u>arrows</u> pointing <u>always</u> from <u>positive</u> to <u>negative</u>.

5) It's vital that you realise that <u>CURRENT IS NOT USED UP</u> as it flows through a circuit. The <u>total current</u> in the circuit is always the <u>same</u>.

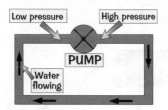

Low pressure   High pressure

PUMP

Water flowing

Current is a bit like water flowing...

The pump drives the <u>water along</u> like a power supply. The water is <u>there</u> at the <u>pump</u> and is <u>still there</u> when it returns to it — and just like the water, electric current in a circuit <u>doesn't get used up</u> either.

Potential Difference Pushes *the Current Around*

1) In a circuit the <u>battery</u> acts like a <u>pump</u> — it provides the driving <u>force</u> to <u>push</u> the charge round the circuit. This driving force is called the <u>potential difference</u>.

2) If you <u>increase</u> the potential difference <u>more current</u> will flow.

3) Different batteries have different potential differences. You can put several batteries together to make a <u>bigger potential difference</u> too.

Potential difference is sometimes called voltage.

Resistance *is How Easily* Electricity Can *Flow*

1) <u>Resistance</u> is anything in a circuit that <u>slows down</u> the flow of current. It is measured in <u>ohms</u> (Ω).

2) You can calculate the <u>resistance</u> of a component by finding the <u>ratio</u> of the <u>potential difference</u> and <u>current</u>. This is just a fancy way of saying:

A <u>component</u> is anything you put in a circuit.

RESISTANCE = POTENTIAL DIFFERENCE ÷ CURRENT

3) This means that as long as the potential difference stays the same, the <u>higher the resistance</u> of a component, the <u>smaller the current</u> through it.

4) Components and materials that electricity can <u>easily</u> travel through are called <u>conductors</u>. <u>Metals</u> are good conductors of electricity (p.42).

Metal strip

Bulb lights

5) <u>Insulators</u> (e.g. wood) are components and materials that <u>don't</u> easily allow electric charges to pass through them.

6) The <u>lower the resistance</u> of a component, the <u>better</u> it is at <u>conducting electricity</u>.

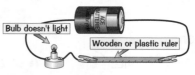

Bulb doesn't light

Wooden or plastic ruler

E.g. bulb A has a resistance of <u>3 Ω</u> and bulb B has a resistance of <u>1.5 Ω</u>. <u>Bulb B</u> has a <u>lower resistance</u>, so bulb B is a <u>better conductor</u> than bulb A.

I power all my cakes with electric currents...

Here we have a page that covers all the <u>really basic stuff</u> about electricity. Comparing electric current with water really does help make things clear, so make sure you get that example clear in your head.

Measuring Current and Potential Difference

Sadly you don't just need to know what <u>current</u> and <u>potential difference</u> are — you need to be able to <u>measure</u> them too. Handily, some clever chaps have made machines to do just that...

Ammeters **Measure** Current

1) <u>Ammeters</u> measure electric <u>current</u>. It's measured in <u>amperes</u> (or amps, A, for short).

2) You measure the current <u>through</u> a circuit by inserting the ammeter <u>into</u> the circuit like this:

3) Remember — current <u>doesn't</u> get used up, so the current through the ammeter is the <u>same</u> as through the bulb.

Voltmeters **Measure** Potential Difference

1) <u>Voltmeters</u> measure <u>potential difference</u> in <u>volts</u> (or V for short).

2) You measure the potential difference <u>across</u> something in the circuit, such as a bulb.

3) To measure the potential difference across a bulb, you'd connect a <u>voltmeter across</u> it like this: <u>Batteries</u> and <u>bulbs</u> have <u>potential difference ratings</u>.

4) A <u>battery</u> potential difference rating tells you the <u>potential difference</u> it will <u>supply</u>.

5) A <u>bulb rating</u> tells you the <u>maximum</u> potential difference that you can <u>safely</u> put across it.

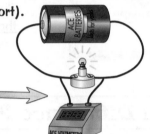

Circuit Diagrams **Represent** Real Circuits

Circuit diagrams are simplified drawings of real circuits. You start at the <u>cell</u> or <u>battery</u> and go round the circuit, <u>putting in the symbol</u> for <u>each component</u>.

Here are the circuit symbols you need to know:

A cell = -⊣|⊦-
(a single energy source)

A battery = -⊣|||⊦-
(a battery is two or more cells put together)

A voltmeter = -Ⓥ-

An ammeter = -Ⓐ-

A bulb = -⊗-

A motor = -Ⓜ-

A switch:

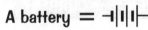 – open = -∘ ⁄ ∘-

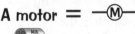 – closed = -∘-∘-

A buzzer = -◡-
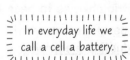

In everyday life we call a cell a battery.

A kilogram of bacon? — Weigh it on the 'ammeter...

You'll need to get your ruler out to draw a nice, neat <u>circuit diagram</u>. They don't look much like the real circuits they show — but they do make it <u>easier</u> to see how everything is connected up. It's a bit like the London Tube map that way. Symbols are also much quicker and easier to draw than the actual components.

Series and Parallel Circuits

The big difference between <u>series</u> and <u>parallel</u> circuits is that in parallel circuits, current can take <u>different routes</u> around the circuit. And the charges don't even need a map or a GPS to do it...

Series Circuits — Current has No Choice of Route

1) In the circuit on the right current flows out of the <u>cell</u>, through the <u>ammeter</u>, the <u>bulbs</u>, then through the other ammeter and the <u>switch</u> and back to the <u>cell</u>. As it passes through, the current gives up some of its <u>energy</u> to the bulbs.

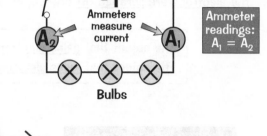

Ammeters measure current

Ammeter readings: $A_1 = A_2$

Bulbs

2) The current is the <u>same anywhere</u> in this circuit as the current has no choice of route. Did I tell you <u>current isn't used up</u> — well don't forget.

> In series circuits the current is either on or off — the switch being open or any other break in the circuit will stop the current flowing everywhere.

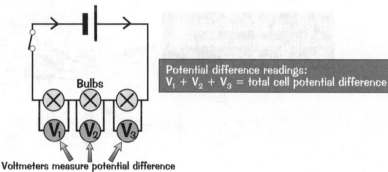

Bulbs

Potential difference readings:
$V_1 + V_2 + V_3$ = total cell potential difference

3) In series circuits, the <u>potential differences</u> across the components <u>add up</u> to the potential difference of the cell (or battery).

Voltmeters measure potential difference

Parallel Circuits — Current has a Choice

1) In the circuit shown, current flows <u>out</u> of the <u>cell</u> and it <u>all</u> flows through the first ammeter A_1. It then has a "choice" of <u>three</u> routes and the current <u>splits</u> down routes <u>1</u>, <u>2</u> and <u>3</u>.

2) The readings of ammeters A_3, A_4 and A_5 will usually be <u>different</u>, depending on the <u>resistances</u> of the components — i.e. the bulbs.

3) The three currents <u>join up</u> again on their way back to the cell. So the readings of $A_3 + A_4 + A_5$ added together will be equal to the reading for current on ammeter A_2 (which will <u>also</u> equal A_1).

4) It's difficult to believe I know, but the current through A_1 is the <u>same</u> as the current through A_2 — the current is <u>NOT USED UP</u>. (I may have told you that once or twice already.)

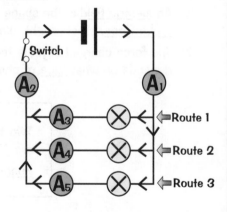

Switch

Route 1
Route 2
Route 3

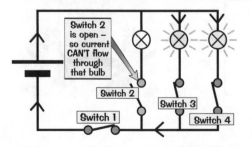

Switch 2 is open – so current CAN'T flow through that bulb

Switch 2
Switch 3
Switch 1
Switch 4

5) Parallel circuits are <u>sensible</u> because part of them can be <u>ON</u> while other bits are <u>OFF</u>. In the circuit here, <u>two</u> bulbs are <u>on</u> and the other one is <u>off</u>.

6) Don't get confused — the <u>potential difference</u> across each bulb in this circuit is <u>equal to</u> the potential difference of the cell.

A series of circuits — well, there are four on this page...

Circuits cause people a lot of gyp, that's for sure. The worst thing about them is that you can't actually <u>see</u> the current flowing, so it's very difficult to appreciate what's <u>going on</u>. Toughsky.

Static Electricity

Right, that's enough on charges flowing about the place. Now let's look at <u>static</u> charges...

Charges Can Build Up When Objects are Rubbed Together

1) <u>Atoms</u> (see page 35) contain <u>positive</u> and <u>negative charges</u>.

2) The <u>negative</u> charges are called <u>electrons</u>. Electrons can <u>move</u>, but <u>positive charges can't</u>.

3) When two insulating objects (see page 95) are <u>rubbed</u> together, the <u>electrons</u> are <u>scraped off</u> one object and <u>left</u> on the other.

> The object that <u>gains electrons</u> becomes <u>negatively</u> charged.
> The object that <u>loses electrons</u> is left with an <u>equal</u> but <u>positive</u> charge.

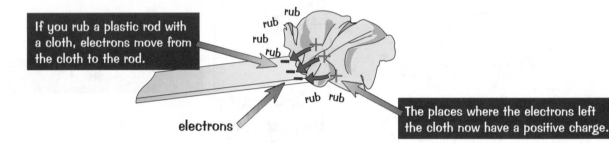

If you rub a plastic rod with a cloth, electrons move from the cloth to the rod.

electrons

The places where the electrons left the cloth now have a positive charge.

All Charged Objects Have an Electric Field Around Them

1) Charged objects <u>don't have to touch</u> each other for them to feel a <u>force</u> from each other.

2) An <u>electric field</u> is the <u>space</u> around a charged object where other charged objects will <u>feel a force</u>. That's right, electric forces can act <u>across a gap</u>. Clever stuff.

3) The force charged objects feel when they come near each other depends on what <u>type</u> of charge they have.

> Two things with <u>OPPOSITE</u> electric charges are <u>ATTRACTED</u> to each other.
> <u>Positive</u> and <u>negative</u> charges attract.

positive charge   negative charge

Two things with the <u>SAME</u> electric charge will <u>REPEL</u> each other.

Electrons put me in a bad mood — they're so negative...

When materials are <u>rubbed together</u> it's only ever the <u>electrons</u> that move — the positive charges never ever get to go anywhere (poor lambs). Static electricity's great fun. You must have tried it — rubbing a <u>balloon</u> against your <u>head</u> and getting your <u>hair</u> to stick up like a crazy scientist's. Your hair sticks up like that because each of your hairs has the <u>same type of charge</u>, so they <u>repel</u> each other. Neat.

Magnets

Electric charges aren't the only things to push and pull each other without touching, magnets can do it too.

Magnets are Surrounded by Fields

1) Bar magnets are (surprisingly enough) magnets that are in the shape of a bar. One end of the bar magnet is called the North pole and the other end is called the South pole.

2) All bar magnets have invisible magnetic fields round them.

3) A magnetic field is a region where magnetic materials (e.g. iron) experience a force.

4) You can draw a magnetic field using lines called magnetic field lines. The magnetic field lines always point from the N-pole to the S-pole.

5) This is what the magnetic field around a bar magnet looks like:

6) You can investigate magnetic fields using either iron filings or a plotting compass...

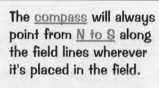

The iron filings will align along the field lines which looks quite jolly.

The compass will always point from N to S along the field lines wherever it's placed in the field.

The field lines (or "lines of force") always point from NORTH to SOUTH.

Opposite Poles Attract — Like Poles Repel

Just like electric charges (see page 98), magnets don't need to touch for there to be a force between them.

Attraction

North poles and South poles are attracted to each other.

Repulsion

If you try and bring two of the same type of magnetic pole together, they repel each other.

The Earth has a Magnetic Field

1) The Earth has a magnetic field. It has a North pole and a South pole, just like a bar magnet.

2) Compasses line up with magnetic fields — so unless you're stood right next to a magnet, they will point to the Earth's magnetic North pole (which handily is very close to the actual North pole).

3) Maps always have an arrow on them showing you which direction is North. This means you can use a map and a compass to find your way.

Magnets are like farmers — surrounded by fields...

Who knew that magnets weren't only fun to play with, but could also stop you getting lost? Make sure you know all the fun magnetic facts on this page. You know the drill by now — learn, cover and scribble.

Electromagnets

Bar magnets stay magnetic all the time. Electromagnets are fancy magnets which you can turn on and off.

A Wire With a Current in it Has a Magnetic Field Round it

1) A current going through a wire causes a magnetic field around the wire.

2) A solenoid is just a long coil of wire. Its magnetic field is the same as that of a bar magnet when a current flows through it.

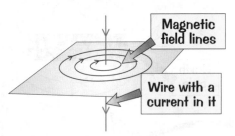
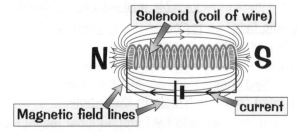

3) Magnets made from a current-carrying wire are called ELECTROMAGNETS.

4) Because you can turn the current on and off, the magnetic field can be turned on and off.

You Can Increase the Strength of an Electromagnet

1) More current in the wire.

2) More turns on the solenoid.

3) A core of soft iron inside the solenoid.

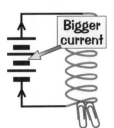
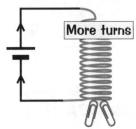
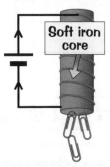

You can't just use any metal to make an electromagnet core. Soft iron has to be used for the core to make it perform as an electromagnet should — i.e. turning on and off when the current is turned on and off.

If a steel core was used, it would stay magnetised after the current was switched off — which would be no good at all.

Electric Motors are Made Using an Electromagnet

1) A simple electric motor is made from a loop of coiled wire in a magnetic field.

2) When current flows through the wire, a magnetic field forms around the wire.

3) Because the wire is already in a magnetic field, there are forces on the loop of the wire. These forces act in opposite directions and cause the loop of wire to turn.

4) Bob's your uncle, you've got a motor.

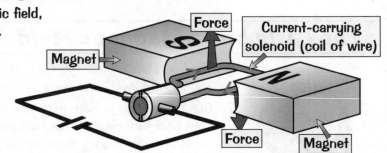

Electromagnets — they're so awesome...

... not only can we use them to make motors, but they're also used in medicine, in scrap yards, in loudspeakers... is there anything they can't do? Well yes, sadly they can't learn this page for you. Make sure you know what an electromagnet is, how to strengthen one and how a simple motor works.

Section Summary

Phew. Electricity and Magnetism — it's no holiday, that's for sure. There are certainly quite a few grisly bits and bobs in this section. There again, life isn't all bad — just look at all these lovely questions I've cooked up for your delight and enjoyment. These are very simple questions which just test how much stuff you've taken on board. They're in the same order as the stuff appears throughout Section 11 — so for any you can't do, just look back, find the answer, and then learn it good and proper for next time.
Yeah that's right, next time — the whole idea of these questions is that you just keep practising them time after time after time — until you can do them all effortlessly.

1) Current is the flow of what?
2) Can current flow in an incomplete circuit?
3) What job does a battery do in a circuit?
4) What is potential difference?
5) What is resistance?
6) What is the difference between a conductor and an insulator?
7)* Component A has a resistance of 1 Ω, Component B has a resistance of 0.5 Ω and Component C has a resistance of 0.01 Ω. Which component is the best electrical conductor?
8) What instrument do we use to measure current? How would you connect it in a circuit?
9) What are the units of current?
10) What instrument do we use to measure potential difference? How would you connect it in a circuit?
11) What are the units of potential difference?
12) What is a circuit diagram? Why don't we draw out the real thing all the time?
13) Sketch the circuit symbol for all of these:
a) a buzzer   b) a bulb   c) a battery   d) a switch (open)   e) a cell
f) an ammeter   g) a voltmeter.
14)* A series circuit contains 3 bulbs. A current of 3 A flows through the first bulb. What current flows through the third bulb?
15) What happens if there is a break in a series circuit?
16) Which type of circuit allows part of the circuit to be switched off?
17) In parallel circuits current has a choice of what?
18) True or false? Adding the current through each branch of a parallel circuit gives you the total current.
19) Explain how a cloth and a plastic rod both become charged when they're rubbed together.
20) Do charged objects need to touch to repel each other?
21) State whether each of these pairs of charged objects will be attracted or repelled by each other.
a) positive and positive   b) negative and positive   c) negative and negative
22) What is a magnetic field? In which direction do field lines always go?
23) Sketch a diagram showing how a plotting compass points around a bar magnet.
24) Name two magnetic poles that will: a) attract each other   b) repel each other.
25) Explain why you can use a compass to navigate.
26) What's a solenoid? What do the field lines around a solenoid look like?
27) What's an electromagnet? List three ways to increase the strength of one.
28) Explain how a simple electric motor works.

* Answers on page 108.

Gravity

It's not <u>magic</u> that keeps your feet on the ground, it's <u>gravity</u>. As Sandra Bullock will tell you.

Gravity is a Force that Attracts All Masses

1) Anything with <u>mass</u> will <u>attract</u> anything else with mass.
   In other words, everything in the Universe is attracted by
   the force of <u>gravity</u> to everything else.
   (But you only <u>notice</u> it when one of the things is really big like a planet.)

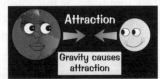

2) The <u>Earth</u> and <u>Moon</u> are <u>attracted by gravity</u> — that's
   what keeps the Moon in its orbit. The <u>Earth</u> and the <u>Sun</u>
   are attracted by an even <u>bigger force</u> of <u>gravity</u>.

3) The <u>more massive</u> the object (or body) — the <u>stronger</u> the force
   of gravity is (so planets with a <u>large mass</u> have <u>high gravity</u>).

4) The <u>further the distance</u> between objects
   — the <u>weaker</u> the gravitational attraction becomes.

Gravity Gives You Weight — But Not Mass

To understand this you must <u>learn all these facts</u> about <u>mass and weight</u>:

1) <u>Mass</u> is just the <u>amount of 'stuff'</u> in an object.
   The mass of an object <u>never changes</u>, no matter where it is in the Universe.

2) <u>Weight</u> is caused by the <u>pull</u> of <u>gravity</u>.

3) An object has the <u>same mass</u> whether it's on <u>Earth</u> or on <u>another planet</u> (or on a <u>star</u>) — but its
   <u>weight</u> will be <u>different</u>. For example, a 1 kg mass will <u>weigh less</u> on <u>Mars</u> (about 3.7 N) than it
   does on <u>Earth</u> (about 10 N), simply because the <u>force</u> of gravity pulling on it is <u>less</u>.

> Weight is a <u>force</u> measured in <u>newtons</u> (N). It's measured using a <u>spring</u> balance or <u>newton meter</u>.
> Mass is <u>not</u> a force. It's measured in <u>kilograms</u> (kg) with a <u>mass</u> balance.

Learn this Important Formula...

$$\text{weight} = \text{mass} \times \text{gravitational field strength}$$

in N   in kg $$W = m \times g$$ in N/kg

1) The letter "g" represents the <u>strength</u> of the gravity and its value is <u>different</u> for <u>different planets</u>.
   <u>On Earth g ≈ 10 N/kg</u>. <u>On Mars</u>, where the gravity is weaker, g is only about <u>3.7 N/kg</u>.

2) This formula is <u>hideously easy</u> to use:

<u>Example:</u> What is the weight, in newtons, of a 5 kg mass, both on Earth and on Mars?

<u>Answer:</u> $W = m \times g$.   On Earth: $W = 5 \times 10 = \underline{50\ N}$ (The weight of the 5 kg mass is 50 N.)
                          On Mars: $W = 5 \times 3.7 = \underline{18.5\ N}$ (The weight of the 5 kg mass is 18.5 N.)

See what I mean? Hideously easy — as long as you've learnt what all the letters mean.

Just make sure you appreciate the gravity of all this...

Remember: <u>weight</u> and <u>mass</u> are <u>NOT the same</u> — they're linked by the <u>formula</u> above. <u>Learn it</u>. You don't
need to know what g is on other planets, but you do need to know that on <u>Earth</u> it's around <u>10 N/kg</u>.

The Sun and Stars

Ahh. This is going to be a <u>nice page</u>, I can tell. Look at all those <u>lovely big pictures</u> for a start.

The Sun is at the Centre of Our Solar System

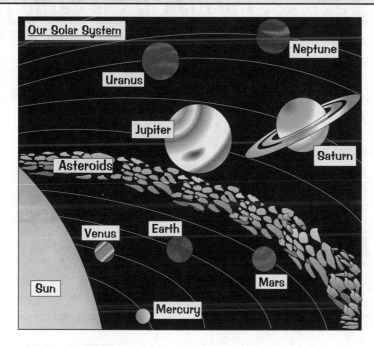

Our Solar System

Neptune
Uranus
Jupiter
Saturn
Asteroids
Venus
Earth
Sun
Mercury
Mars

1) A <u>planet</u> is something which <u>orbits</u> around a <u>star</u>.

2) The <u>Sun</u> (at the <u>centre</u> of our <u>Solar System</u>) is a <u>star</u>. The <u>Earth</u> is one of <u>eight</u> planets which orbit the Sun.

3) The Sun is really <u>huge</u> and has a big <u>mass</u> — so its <u>gravity</u> is really <u>strong</u>. The pull from the Sun's gravity is what keeps all the planets in their <u>orbits</u>.

4) The planets all move in <u>elliptical orbits</u> (elongated circles).

5) Planets <u>don't</u> give out light but the <u>Sun</u> and other <u>stars</u> do.

6) The Sun gives out a <u>massive</u> amount of <u>energy</u> which is transferred by <u>light</u>.

Beyond the Solar System

1) A <u>galaxy</u> is a <u>large collection</u> of <u>stars</u>. The <u>Universe</u> is made up of <u>billions</u> of galaxies.

2) Most of the stars you see at night are in our own <u>galaxy</u> — the <u>Milky Way</u>. The other galaxies are all <u>so far away</u> they just look like <u>small fuzzy stars</u>.

You are here in the Milky Way

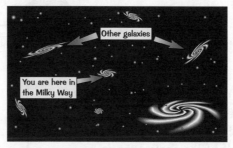

Other galaxies
You are here in the Milky Way

North Pole

3) There are <u>billions</u> of stars in our galaxy, including the <u>Sun</u>.

4) Other stars in our galaxy include the <u>North star</u> or <u>Pole star</u> (which appears in the sky above the <u>North Pole</u>) and <u>Proxima Centauri</u> (our <u>nearest star</u> after the Sun).

A Light Year is a Unit of Distance

1) A light year is <u>how far light travels</u> in <u>one year</u>.

2) It's used for measuring <u>huge distances</u> between objects — like the distances you find in <u>space</u>. E.g. <u>Proxima Centauri</u> is about <u>4 light years</u> away, which means it takes light from the star 4 years to <u>reach Earth</u>.

4 light years

Proxima Centauri

Earth

1 light year ≈ 9.5 million million km (a <u>long</u> way)

Galaxies, the Milky Way — physicists must like chocolate...

See, I told you this would be a nice page. And an easy one at that. So there's <u>absolutely no excuse</u> for <u>not learning everything</u> on it. You know the score by now. <u>Cover</u> up the page and <u>scribble</u> down everything you can remember. Then <u>check back</u> over the page to see what you missed. Then <u>try again</u>, till you <u>get it all</u>.

Day and Night and the Four Seasons

In years to come, this stuff will come up in a <u>quiz</u> and you'll be able to <u>wow</u> your teammates with the answer. You also need to know it for <u>KS3 Science</u>. So get cracking...

Day **and** *Night* **are** *Due to the* **Steady Rotation** *of* **The** *Earth*

1) The Earth does <u>one complete rotation</u> in <u>24 hours</u>. That's what a <u>day</u> actually is — <u>one complete rotation of the Earth</u> about its axis.

2) The Sun doesn't really move, so as the Earth rotates, any place on its surface (like England, say) will <u>sometimes face the Sun</u> (<u>day time</u>) and other times <u>face away</u> into dark space (<u>night time</u>).

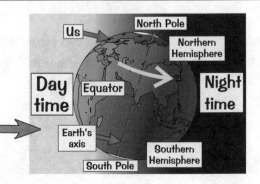

The *Seasons* **are** *Caused* **by the** *Earth's Tilt*

1) The Earth takes <u>365 ¼ days</u> to <u>orbit once</u> around the Sun. That's one year of course. (The extra ¼ day is sorted out every <u>leap year</u>.) Each year has <u>four seasons</u>.

2) The seasons are caused by the <u>tilt</u> of the <u>Earth's axis</u>.

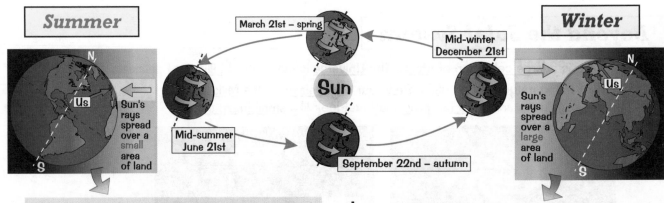

Summer

Sun's rays spread over a small area of land

Winter

Sun's rays spread over a large area of land

March 21st – spring

Mid-winter December 21st

Sun

Mid-summer June 21st

September 22nd – autumn

1) When it's summer in the UK, the <u>northern hemisphere</u> (everything above the equator) is tilted <u>towards</u> the Sun.

2) The northern half of the Earth spends <u>more time in sunlight</u> than it does in darkness, i.e. <u>days are longer</u> than nights. Longer days mean <u>more hours of sunshine</u> — so the land <u>heats up</u>.

3) Not only that, but the Sun's rays cover a <u>small area</u> of land. This means that the <u>radiation</u> is <u>focused</u> on a small area. So it gets <u>warm</u> (see p.67) and we have summer — hoorah.

1) When it's winter in the UK, the northern hemisphere is tilted <u>away</u> from the Sun.

2) The north now spends <u>less time in sunlight</u> so <u>days are shorter</u> than nights.

3) Also, the Sun's rays cover a <u>larger area</u> of land so the radiation is <u>more spread out</u>. So it gets <u>colder</u> and we have <u>winter</u>.

When it's <u>summer</u> in the <u>northern hemisphere</u>, it's <u>winter</u> in the <u>southern hemisphere</u> — and vice versa.

Phew — I feel quite giddy now...

Well this is all very jolly. Not like usual boring old science at all. This is really interesting. OK. Fairly interesting. Anyway, learn the <u>headings</u>, the <u>diagrams</u>, the <u>details</u> — and <u>scribble</u>. Make sure you know how the tilt of the Earth's axis gives us <u>warm summers</u> (yeah, right) and <u>cold winters</u>.

Section Summary

Section 12 only has three pages of information — not much really, considering it deals with the whole Universe. It's amazing just how many people go their whole lives and never really know the answers to all those burning questions, like what is gravity? Or why are the days longer in summer than in winter? Make sure you learn all the burning answers now...

1) What is gravity?

2) Which is stronger, the gravitational attraction between the Moon and the Earth or the Sun and the Earth?

3) How does the mass of a planet affect its gravitational field strength?

4) What is the difference between weight and mass?

5) What is weight measured in? What is mass measured in?

6) In the formula $W = m \times g$, what does 'g' stand for?

7) On Earth, what does 'g' equal?

8)*On Jupiter, g = 25 N/kg. What would a 5 kg mass weigh on Jupiter?
Remember to include the correct unit in your answer.

9) What is at the centre of our Solar System?

10) How are all the planets kept in orbit around the Sun?

11) What is a galaxy?

12) What is the name of our galaxy?

13) Apart from the Sun, name one other star in our galaxy.

14) What is a light year?

15) How long does it take for the Earth to complete one full rotation on its own axis?

16) Explain what "day time" and "night time" actually are.

17) Do all places on the Earth have "day time" at the same time?

18) How long does it take for the Earth to complete one full orbit around the Sun?

19) How many seasons are there?

20) Why are days longer than nights in summer?

21) Give two reasons why it's (supposedly) hotter in Britain in summer than winter.

22) When it's summer in the northern hemisphere, what season is it in the southern hemisphere? Explain why.

* Answers on page 108.

Index

Index

Index and Answers

Answers to Top Tip Questions

p.24 a) Fewer otters means more pike, which will eat more water beetles.

b) More pike would mean fewer perch, which could mean fewer water beetles get eaten.

p.38 a) sodium fluoride b) iodine c) calcium sulfate

p.51 $2Na + Cl_2 \longrightarrow 2NaCl$

p.82 1) Balanced. 2) Balanced. 3) Unbalanced — right side down.

4) Unbalanced — left side down. 5) Balanced. 6) Balanced.

Answers to Selected Section Summary Questions

Section Summary 2 — Page 18

6) Daily basic energy requirement = 5.4 × 24 hours × body mass (kg) = 5.4 × 24 × 54 = 6998.4 kJ

19) Size of moment = force × perpendicular distance = 8 × 0.2 = 1.6 Nm.

Force applied by the muscle = moment ÷ perpendicular distance = 1.6 ÷ 0.04 = 40 N.

Section Summary 5 — Page 47

23) a) magnesium oxide b) calcium oxide c) sodium chloride d) calcium carbonate e) copper sulfate

24) a) sodium chloride b) magnesium chloride c) magnesium carbonate

Section Summary 6 — Page 58

8) exothermic   **9)** endothermic   **13)** $S + O_2 \longrightarrow SO_2$ **14)** $2Ca + O_2 \longrightarrow 2CaO$

Section Summary 8 — Page 76

6) Energy transferred (J) = force (N) × distance moved (m) = 2000 × 10 = 20 000 J = 20 kJ

19) Energy transferred = power (kW) × time (h) = 1.5 kW × 0.5 h = 0.75 kWh

21) Cost = energy transferred in kWh × price per kWh = 298.2 × 15 = 4473p = £44.73

23) The 300 W device (it has a higher power rating).

27) 50 g (the amount of substance is the same before and after).

Section Summary 9 — Page 85

3) s = d/t = 5 ÷ 2 = 2.5 m/s.   **4)** s = d/t = 100 ÷ 20 = 5 m/s. Yes.

5) s = d/t so d = s × t = 40 × 0.25 = 10 miles. (15 minutes = 0.25 hours).

27) 50 × d = 100 so d = 2 m.   **31)** 200/2 = 100 N/m^2 (or 100 Pa).

Section Summary 11 — Page 101

7) Component C (it has the smallest resistance).   **14)** 3 A

Section Summary 12 — Page 105

8) W = m × g = 5 × 25 = 125 N